COSMOLOGY
revisited
Book 2

Copyright 2015
George W. Harper

**DEDICATED
TO THOSE WHO CAN
THINK OUTSIDE THE BOX**

PROLOGUE

Those who have ploughed through the first book without regurgitating will be relieved to hear that I am exercising more restraint in my follow-up chapters here. But there will no doubt be occasions when temptation overwhelms me. For this I apologize in advance; but cosmology is too serious an affair to be taken seriously. The very idea of humans making value judgments about the universe is akin to the idea of ant scientists making value judgments about New York City.

The thought of arrogant humans flapping their arms and crowing about our achievement in comprehending the universe is ludicrous... and the more the pompous and pretentious the learned professor or preacher who declaims about it, the more hilarious the ensuing oratory becomes.

Enough said.

Next comes a personal apology for the type face used here. I am 88 years old and nearly blind. In brief, this type size and intensity is the least I can manage in order to review where I left off a day or so earlier or proof-read the text as a whole one or two months before that.

Again; enough said.

Now for a few preliminary remarks about the little item we call 'cosmology'.

Cosmology and astronomy are in reality two distinct but frequently overlapping sciences. In gist, the astronomer looks through the telescope and manipulates the tools while the task of the cosmologist is to tell him what he sees.

Book 1 centered almost exclusively on the cosmology of the universe, its origins from the primal ylem and progress up to the evolution of galaxies. This meant having to address wholly new types of problems and argue about ideas largely ignored by cosmologists and therefore required ideas which entailed the breaking of new ground almost every step of the way.

The presentation in Book 1 was

largely philosophical. Of course, I could have resorted to academic obfuscation by strewing Cyrillic, Aramaic, Greek and other forms of notation all over the landscape whereby I could awe the yokels into blind submission to my genius but somehow or another this ploy fails to suit me. Rather than inducing awe I would much prefer to communicate understanding.

In book 2 we possess a remarkably precise and copious mass of firm data; in which the astronomers may take legitimate pride. But this does not mean we are home free, with no little mysteries or unexplored paradoxes lurking beneath the surface. What it does mean is that we are free to focus on any given problem without risk of getting mired in a needless mass of superfluous digressions.

Nothing in this alters my self-imposed task of developing a coherent overview of both the megaverse and the universe. Book 1 disposed of the megaverse. Book 2 should strive for the same level of coherence; which implies an analytical approach based on existing data.

Over the years I have had about a dozen articles involving elements of significance in this aspect of cosmology published in
various magazines. I do not take pride in all of them, but several broke new ground. When appropriate, I shall republish these articles with emendations as needed... specifically including those portions where I was in error. This may make for interesting reading at a time when I am offering my *mea culpas* for some idiocy or another.

On this note my prologue is complete.

CHAPTER I

WHOLLY HOLEY BUT NOT HOLY

As preface to Book 1 I cited Artemus Ward's little dictum which says that it is not what you don't know that's half so apt to hurt you as the things you do know that aren't true. Perhaps nowhere is this more accurate than in our comprehension of the galaxy.

Shall we be systematic rather than hurling accusations or engaging in piecemeal dogmatism? This sounds like a good idea so perhaps we should try it for a change.

Start with Schwarzschild. During the 1920's Schwarzschild arrived at the theory of the black hole. It was perfectly straightforward and derives directly from Newton, and before him from Euclid. In brief, it was the very essence of simplicity. For the benefit of those unacquainted with the niceties of astrophysics please think back to old Euclid and see what happens.

Consider the little progression:

.=0, 1=1, 2=1/4, 3=1/9, 4=1/16, 5=1/25, 6=1/36, 7=1/49, 8=1/64, etc.

Now translate this into plain English.

We start with a point. Think of it as the dead center of our Earth. It is surrounded by tonnes of rock, much of it in the form of molten magma. The gravity in every direction equals the gravity in every opposite direction, thereby cancelling itself out. If we could somehow insert a heatproof, structurally rigid capsule into this center and set about exploring the core any occupants of the capsule would find themselves floating in midair just as the astronauts in spacelab do. The competing gravitations coming from every direction would simply cancel each other out.

So '.=0'. Easy enough.

The next point in our little excursion lifts us to the surface of our Earth. Call it 6,000 km and be close enough for any normal purpose. At this point the acceleration of gravity works out to be around 982 cm/sec^2. Call it 1,000 cm/sec^2 and be done.

So 1=1. The distance from the core is taken as 1 and at this distance the acceleration of gravity is defined as 1.0. Again the concept is simple enough. At twice the distance from the origin the attraction of gravity is 1/4 the surface gravity and at 3 Earth radii it is 1/9, and at 4 radii it is 1/16, etc.

This is the inverse square law as expressed by Newton, who copied it from Euclid, who copied it from Pythagoras, who may well have copied it from Nabu Rimanni, who probably copied it from some much older source, perhaps some Sumerian astronomer or the Egyptian Imhotep sometime around 4,000 b.c.e. In short, the inverse square rule dates back to the earliest days of civilization.

Now comes the little joker in the woodpile. The inverse square law works both ways. If the gravity at 2 radii is 1/4 that of Earth's surface then if we could compact the Earth to a centimeter of its present size the gravity must be 4 times that at the present surface, i.e. 4,000 cm/sec^2.

This is technically somewhat in

error since there would be competing forces pulling in opposing directions but the concept is valid.

Given this, an astronaut skittering along over Earth's old surface and measuring the gravitational force at 5,000 km would still detect a gravitational attraction of 1,000 cm/sec^2 so nothing would be changed so far as the system as a whole is concerned.

Now according to theory, if the gravitational attraction of a mass is equal to or greater than 297,000 km/sec that object simply disappears from view. Given that the speed of light is 297,000 km/sec the gravity holds the light in thrall. Not even light can escape because of the enormous gravity. Given the mass implied by the gravitation we learn that a single cubic centimeter of iron might well be defined as infinite... or at the least as several hundred, or even thousand, metric tons.

Again according to current theory about the only way a hole can be detected is through discovery of Cerenkov radiation in the region immediately surrounding the hole. This type of radiation is indicative of gaseous material which has been accelerated to a speed greater than the speed of light in the medium through which it is moving. We routinely detect this radiation in the regions surrounding the cores of nuclear reactors. The color of this Cerenkov radiation is a pale, virtually invisible blue of a hue similar to that of a cloudless sky on a warm summer day and is often described as 'washed out blue'; which may suggest that our blue skies are a form of Cerenkov radiation created by the interaction of sunlight with our atmospheric gasses.

Apart from this there is only one means of detecting a black hole, i.e., via the gravitational warping of space. For example, the trajectory of light passing to one side of a hole will be arced around the hole so its track will be skewed. I suppose that in theory there might be occasions where a beam of light is given a hyperbolic trajectory where it is reflected back virtually to its origin, with an even more improbable solution making a beam of light shift into orbit around a hole... in which event we would never know it was there unless we were inserted into the same orbit!

In an exceedingly unlikely case we might even arrive at a figure 8 orbit for light trapped between a pair of holes in binary orbits about each other, but this is getting pretty exotic.

Of course this scenario is dismissed in Book 1, but here I am specifically pursuing the orthodox models and pointing to a few of the numerous problems available. And this is only the tip of the iceberg.

Photographs delving into the deepest depths of space reveal hundreds or even thousands of spiral galaxies presumed to be relict of the earliest days of the universe.

This may be somewhat misleading because spiral arms are generally more luminous than the stars of routine globular galaxies; which may not be bright enough to be singled out at so great a distance. Even so, spirals are among the most ancient of galactic systems and date back about 13.8

billion years.

In effect, this argues that the spiral types are pretty much typical of galaxies from their emergence as star systems. But this cannot be true of our galaxy; which consists of a collection of otherwise tightly wound spiral arms comprised of type 1 stars superimposed atop a conventional globular array of type 2 stars. In other words, first came a galaxy packed with metal poor type 2 stars. Then came an odd event which wound up fusing the higher elements into existence as type 1 spiral stars, appearing in the form of a plane wound around the galactic equator.

When we try to reconstruct the time line involved here we find yet another problem. For the nonce I use a conventional 13.8 billion year model as the age of our universe.

Our planet has been fairly accurately dated as being some 4.5 billion years old. But this may well be an exaggeration. Stop to consider some of the implications here. Meteoroids have also been dated to around 4.5 billion years. I may be slightly in error on this but I take it as indicative of a prior stellar event which fused lesser atoms into heavier, metallic varieties. If so, when we speak of a 4.5 billion year old Earth we are really speaking of the moment when the primal event which ultimately coalesced into the planet finally got together and began the process.

There is indisputable support for this conclusion. Earth, along with Mercury, Venus and Mars provide ample evidence of iron, nickel and other heavy elements, and the presence of strong magnetic fields around Jupiter, Saturn and Neptune suggests a metallic presence in them as well; though hydrogen is classed as a metal so it would be dangerous to place too much emphasis on this line of reasoning. But in addition, the nickel/iron meteorite is well known, albeit not as abundantly as ordinary meteorites.

Given the facts that type 2 stars are noteworthy for their lack of iron, and type 1 stars are specifically characteristic of spiral arms we may take it as proven that our origin is an outgrowth of material cast off from the galactic core in the form of spiral arm elements. This must be a secondary development taken after a series of intervening events, but it has to be essentially correct.

Now consider the postulation of a black hole nesting in the heart of the galaxy.

At first glance this would appear to be a contradiction to all existing theory. The spiral arms proclaim a titanic outpouring of raw energy streaming *away* from the core, not into it as we might expect! Core luminosity shows no sign of systemic absorption of luminosity into a hole. How then did the core manufacture all the iron we find in the spirals? And, most damning of all, the spiral arms must have been in existence blazing incandescently for well over 5 billion years; possibly even around 13 billion. Where did all that energy come from? Equally curious, if nothing can escape from a black hole, how did all that mass manage to get out: through a back door perhaps?

It seems conclusive, there cannot

be a black hole doing all the work! You cannot devise a rational system where nothing can escape while it is simultaneously decanting tens of millions of stars laden with multi-bevatonnes of irons and other higher elements in a constant stream lasting billions of years.

The statement that 'nothing can escape' means nothing can escape. So how can we resolve the paradox?

But let us not be too hasty here. Things are not always as cut and dried as they seem and there may yet be some way of preserving our black hole intact, thereby soothing the injured feelings of countless pontificators.

In Book 1 I introduced Olber's paradox and pointed out that the cores of galaxies were the only places where it might be a significant factor. It is now time to examine that prospect in some detail.

The exact dimensions of the galactic core cannot be determined with any precision and there may be a noteworthy equatorial bulge to make matters even more imprecise. Even so, a diameter of 5,000 light years should provide a useful working figure.

As for a hole itself, it is difficult to assign it a maximum diameter, but a neutron star has a theoretical diameter running somewhere in the vicinity of 50 km. Based on this a hole may be assigned a diameter running roughly 5 km. Given this, a hole seven times as massive ought to have a diameter in the vicinity of 15 km while a 50 stellar mass hole might be around 30 km.

This is probably incorrect and I shall have occasion to modify it later but this will do for starters. The total number of stars in the galaxy has been variously estimated but usually comes out as somewhere between 100 and 150 billion stars, though there are more extreme estimates make it to be as much as 200 billion! But these are all guesses. You pay your money and take your pick. My guess is that the one to two hundred million figure I have seen quoted is restricted to the spiral belt arms while the other estimates are efforts to count all the stars in the galaxy. Either way, I am opting to base my calculations on a galactic population of 150 billion stars, with perhaps a few billion extras left over as too small to be seen or lost behind other stars or behind clouds, etc. Now throw Olber's paradox into the mix and see what happens.

No matter what number we use the combined temperature coming from all of these stars when it arrives at the core of our galaxy must exceed $1,000°c$! Translated, the mass implied by these $1,000°$ will represent several thousand metric kilotonnes per standard 24 hour day. And please bear in mind that there is no night darkening around the core. It is perpetually blazing hot at every point there.

In other words, a black hole resident at the core of our galaxy will continue gaining in mass even if no particulate matter is absorbed. Since any black hole lurking at the core must have existed at least a dozen billion years it has probably doubled or tripled its mass since it formed; and that derives from this single source alone. Infalling matter will likely have added a further doubling.

I acknowledge the imprecision

inherent in my employment of all these 'perhaps', 'probables', and 'implieds' here, but I am only being honest. I could write some fanciful equations guaranteed to impress the yokels, but if each of the terms is translated into the national patois they will amount to nothing more than integrations with a stated range from zero to infinity, which means that no matter what number you crank in it only defines a solitary possibility out of an infinite number of possibilities. Which translates to mean the researcher can continue manipulating the parameters until he arrives at the answer he wants to get!

Big deal!

Go to the next little conundrum. Note that we are back to the "fabric of space" mythology. Ask how it is possible for a nothingness possess a fabric? Worse yet, ask how it is possible for this nothingness to manufacture energy when it lacks any hint of energy itself? And where are all these mysterious dimensions hiding out?

This is pure theology. It is not

science. Fundamentalist preachers smile in smug contentment at the preachments of these professors of knowledge, but the facts remain intractable. You cannot get something from nothing. Newton pointed out that for every action there is an equal and opposite reaction. Our modern mythologists (call them 'Lysenkoists and be done with it) will cite Newton whenever it is convenient, but where do all the oddments supposedly pouring in from this other dimension go to balance out the equation?

It is a puzzlement which can only be resolved if we postulate that the black hole must contain the aggregate mass of tens of billions of stars. Assume this and we must wonder why the gravitational constant from the hole can be low enough to avoid a prompt wholesale absorption of the other core stars which have persisted to this day without being detectable by our observational techniques. The only way to do this is by assuming the annihilation of two or three stars per day for a period of at least 6 billion years with a consequent production of 'death rales' from some 1.1 trillion years past.

Even more to the point, if gravity is purely notional, how can a black hole exert any energy on its surroundings? If even light cannot escape how can gravity escape to attract outside masses?

In other words, without taking refuge in exotic clandestine factors we are left with an impossibility.

Now suppose we take a look at how this scenario plays out when we use the model developed in Book 1.

There are no mystical dimensions nor are there uncanny attributes lurking in nothingness. Space may be occupied but it is not structured. Gravity is not notional but real and mass relates to gravity in much the same way as water relates to ice.

Continue with the same line of reasoning.

Being the tiniest feature of the megaverse (which includes our universe) any attempt to study the phote directly is destined to fail. It would be very much like using a surgeon's scalpel to dissect a virus or the Hubble telescope to study an amoeba. But we can employ indirect techniques to learn more about photes... though this must be done carefully lest we insensibly drift across the line and devolve into mysticism and theology.

Begin by observing that Schwarzschild worked under the assumption that light is a wave where we are working on the assumption that the phote is a slightly torqued particle of energy which can attach itself weakly to the tail of another phote just as magnets latch onto the oppositely charged tails of other photes while rejecting those of the same polarity. This torqueing is not a megaversal bias but a consequence of the polarization of the vortex created when our particular universe formed.

Now we shift gears to consider the light from the ambient galaxy as it strikes the hole. Each beam consists of connected streams of photes plunging into a dense soup of photes, packed together so tightly they could almost be regarded as a solid mass, but perhaps thinking of them as a different sort of plasma would be more accurate.

If matters stopped there the problem would be intractable and we would be left with nothing better to do than revert to mysticism along the lines of current orthodoxy. But there may be an escape hatch available.

Schwarzschild premised his mathematical calculations on the wave lengths of light. But in our model wave length is notional rather than real. Reverting to our analogy in Book 1, legs are real, but the footsteps are notional, i.e., there cannot be a footstep without feet. Take away feet and there can be no footsteps, but there can easily be feet without footsteps. It is a crude analogy but it a legitimate one.

Unfortunately, we are still at an impasse. If we cannot resolve it we may have to become theologians and start speaking of a 'fabric' of space or similar nonsense.

Now I must make an apology to myself. When I retraced my steps and considered the matter from the beginning it dawned on me that I had made precisely the same mistake I have accused the astrophysical community of committing.

Has anyone spotted it? I dare say most have not, but I hope at least a handful have.

In the simplest terms, I allowed myself to become so wrapped up in my black holes I failed to consider all of the consequences of my own earlier conclusions. And by failing here I arrived at an impasse. Now I must atone for my error.

Our universe is mechanically similar to a smoke ring passing through the air. The underlying ylem is essentially a vast, infinite sea of elemental photes which are darting about aimlessly in all directions, but are tracking about according to the laws of Euclidian straight line geometry. By contrast, the photes which are components of the smoke ring are tracking along the curved geometry imposed by the torus. In short, I was seeking to link two competing Cantorian C sets. In effect, I was trying to square the circle; a tactic foredoomed to failure.

Other than this incompatibility between C sets there is no difference between the two photes.

Now ask what happens when the curved geometry photes are captured within the core of a hole?

This leads to an interesting line of speculation. The pea soup plasma implicit within a hole is turbulently chaotic. I am no doubt incorrect in the detailed analysis but we have the six degrees of freedom inherent in the creation of an explosion in space together with the orthodox straight line geometry already mentioned for the ylem. So the pea soup approach to the heart of a hole necessarily requires photes occupying seven different configurations. Six of these will be consonant with conditions within one or another torus while the seventh... the straight line solution... is alien to all of these!

Now what?

In the crudest sense of the term, one seventh of these photes will not belong in this or any other toroidal universe, ergo they fall out of the hole and rejoin the primal chaos from which they sprang!

Can this be correct?

Of course it is not. I am speaking of a pea soup mix which is in violent motion and collisions are incessant. Warpages will be commonplace, with trillions occurring every second. I doubt whether even the most devoted practitioner of chaos theory can succeed in digesting this much chaos in one gulp, but it would be interesting to see him try. All that is certain is that there will be a torrential leakage of photes out of our universe and back into ylem from which it sprang. As an aside here, this may be what Hawking was deducing when he pointed to 'worm-holing' as a means of escape from a hole. His equations certainly allow for this interpretation.

Assisting this is the effective zero gravity existing at the exact center of the hole. It constitutes an escape hatch which permits wholesale phote seepage directly from the core.

Now comes the question of the effect of this seepage from the hole back into the ylem? And this is the colour of a different horse. The hole is severely limited in size and the infusion of so great a mass of dark matter photes streaming into the ylem from so limited an area will contribute to the mass of dark matter in the region enclosing the hole. What would normally be a gentle stream in a given region becomes a veritable Niagara of dark matter which permeates the entire core region of the galaxy.

In the process of escaping from the hole this dark matter manufactures an overwhelming jet action propelling nearby streams of light and matter toward the hole.

Describing it a bit differently, Back in the ylem there is a massive infusion of photes erupting into a tiny point and streaming out in all directions. It is precisely the same as the unwaved dark matter discussed in Book 1 and depicted as moving according to a lineal geometry while our universe is a toroidal smoke ring passing through it.

This would imply a wholesale

infusion of photes busily manufacturing overwhelming magnetic fields and the existence of an enormous spallation region capable of transmuting lighter elements into transferric elements.

Since the dark matter photes are already streaming away from the hole they manufacture a magnetohydronamic field which converts into one or another of the conventional spiral patterns we see through our telescopes; either that or a polarized ejection will develop into a pair of ejected spikes emerging at right angles to the equatorial plane of the galaxy.

A minor subset to this scenario points to the possibility that there is more than one hole in the galactic core. Dr. Frebel has suggested an extra-galactic origin for globular clusters. These mini-galactic clusters typically consist primarily of 60,000 or so primarily type 2 metal poor stars, with only occasional type 1 members. As a general rule globular clusters are found in the outer regions of the galaxy where type 1 stars are not seen so the occasional type 1 stars are anomalous and deserve explanation.

Here the answer is fairly simple. Clusters generally have highly eccentric orbits; orbits which see them approach the galactic core every few tens of millions of years or so, then swing back toward the galactic fringe. This suggests a possibility that sometime over the past nine or ten billion years a globular cluster slammed into the galactic core, manufacturing a few dozen holes in its progress. Other globulars may have missed the core but acquired an occasional type 1 star during their passage through the spiral arms.

This prospect changes nothing but it offers an alternative which should be kept in mind.

A more remote possibility does not require a globular passage. The galactic core is fairly densely populated with stars and a mean distance of less than one light year is not unreasonable. While actual collisions would be rare near misses would be commonplace. In such instances there would be a transfer of mass, with the larger star drawing mass away from the smaller. This would not only tend to augment the mass of an existing hole but would promote the development of new holes within the core.

If all of this seems confusing this is as it should be. As remarked elsewhere, the universe is a messy system and the antique notion that "From a clockwork we may deduce the existence of a clockmaker" has little to recommend it. It is the product arising from the closed minds of theologians and has no relation to science.

The universe is as it is, not as we might wish it to be.

By no means does this exhaust the subject of holes. There is another intriguing line of thought which merits attention. Holes ought to develop in two entirely distinct environments which display different characteristics. There will be those that are produced in or in close conjunction with galactic cores as well another class which occur around the fringes of the galaxy and are relict of the first or second stellar generations.

It is reasonable to expect the existence of considerably more than 50,000 of these holes strewn about our galaxy, but thus far we have found evidence of perhaps a half dozen... and the verdict is still inconclusive on most of these. They are probably holes but we cannot be entirely certain.

This brings us back to the old drawing board again. What can we rationally expect if we back off and ask what happens when a 75 or 80 solar mass star which has been burning itself up for a million or so years finally runs out of fuel? Does it simply gutter out or does it indulge in a bit of fireworks?

The picture is cut and dried so far as it goes. But it does not go far enough. An extremely low mass star; say one only slightly larger than Jupiter, simply subsides into a brown body which may or may not be satellite to a larger star. Since Jupiter is known to be generating more heat than it receives from the sun we would not be too far of the mark by designating it as a Type I (infrared) star and think of it as the tamest sort of burned out star. An alternate view might call it merely an oversized planet which failed to make it to stardom; a sort of Hollywood wannabee who lacked the talent to outgrow the "B" movie roles.

Larger, better developed stars, wound up following a series of well understood tracks, with the intermediate sized stars taking up residence as white dwarfs, neutron stars, pulsars after penultimate stages as variables having decreasing periodicity *en route* to a nova or supernova event.

The documentation here is essentially cut and dried. It is interesting but it represents a well understood niche in the field of cosmology which we need not discuss here.

The same cannot be said of holes. Only Hawking has bothered with the question of what goes on in the interior of one of these holes. Unfortunately, he began with an uncritical acceptance of existing orthodoxy.

Being unorthodox by nature I dare

to venture into realms where angels fear to tread.

Galactic core holes have virtually unlimited fuel sources. They can pass Niagaras of photes back into the ylem for aeons of millennia because they have outside sources feeding them. Holes residing outside the core and far out among the halo stars lack such an advantage. And this raises intriguing questions.

There are three ways of accounting for the apparent lack of holes. For one, our estimate of the number of holes may be flawed and they simply are not there. A second alternative argues that they have outgassed and wormholed themselves into oblivion. The third argument requires a rethinking of wormhole mechanics. Any one, perhaps all of these alternatives may be correct but the third alternative seems most interesting.

Consider a first or second generation giant population 2 star of the sort postulated as typifying the initial stars forming out in the torus. This star is out on the fringe of what was ultimately to become the halo field. It has been blazing merrily for 60,000 or so years... which is about par for super giants. Now it undergoes a tremendous expansion before finally collapsing into a supernova, throwing off 80 or 90 percent of its mass in the process.

Once again old Newton sneaks back into the picture. For every action there is an equal and opposite reaction. The explosion which ejects so much matter into the depths of space also compresses the remaining 10 to 20 percent of the star into a massive black hole having a diameter of perhaps one kilometer!

Think on this for a minute. What started out as a super-giant star massing between 60 and 100 solar masses casts off the equivalent of between 48 and 90 suns into the void while compacting 10 to 12 suns into a turbulent little bubble boasting a diameter only slightly more than a kilometer!

That is a mighty tight squeeze. As a comparison expressed in human terms, if you could fit the Empire State building into a thimble you still need an electron microscope to see it!

We poor mortals are very hard pressed to contemplate so outrageous a scenario, but it is true nonetheless. We might as well get used to it.

So we now we have our hole far out among the halo stars. What is the next step?

We still have the outpouring of lineal photes streaming back into the ylem as dark matter interacting to provide a soup of magnetohydromagnetic/mass/gravity permeating the ylem space at distances ranging out perhaps a full light year before it dissipates entirely, but the region is lacking the density of matter characteristic of core regions. Some spallation occurs, but it is pretty much a minor affair. What matters is the developing condition at the core of the hole.

Here we have a situation where much more mass/gravity is being expelled from our universe than is being replaced from the surrounding space.

Now we ask what must happen

when the loss of mass from the hole leads to a situation where it can again radiate light and is thus no longer a 'black' hole? What does it look like?

One thing we can be certain of... It will be hot as a pistol! In astronomy we learned that red stars are reasonably cool, at least at the visible surface. Hotter stars shift toward yellowish, hotter yet to white and the hottest stars are blue. So these reradiating ex-holes will glow well into the blue end of the spectrum.

How to distinguish this blueness from Doppler shifting in the case of stars moving toward us is fairly simple. Any stars in the galaxy will be limited in their speed of approach or recession. This sets a limit on any blue shifting and such blueness can be discarded as an artifact of motion.

The second element in the equation will be the faintness of the star. It may be enormously hot but it is tiny; just a few km in diameter. It is virtually a point source.

This brings us to a third element in our analysis. We have a hot, blue star. What more do we need?

For another factor we may cite expansion. The erstwhile hole is losing its compression and the result is not unlike watching a fat man remove his belt. A lot of stomach pops out. A hole which started out with a diameter of one kilometer might grow to 1.2 km, then to 1.5 or even 2.0 km.

In the process there may be sporadic outgassings where the former hole bursts out with surface eruptions which rapidly cool to merely white or even yellow. Ultimately the ex-hole may settle down to join the ranks of neutron stars, possibly as stellar bursters, or even graduate to become white dwarves if it is really ambitious.

Regardless of this last bit of speculation it ought to be clear enough by now: the black hole is far more complex than current theory contemplates.

In essence, the absurdity of theorizing a hole from which even light cannot escape while a notional gravity inexorably drags material into its gaping maw and somehow spits out millions of stars over billions of years becomes ridiculous on the face of it!

How could this have happened? The answer is simplicity in its symmetry. It is a logical outgrowth of our teaching and learning system; a product of over specialization. One pedant looks to the gravitation of a hole and works from there. Another looks to the spiral arms and uses that as his starting point. Each has his specialty and neither inclines to poach on the other's turf.

How to resolve the dilemma? My only suggestion is to create a new level of science education. Call it a 'generalist' class where the primary function of such pedants is to seek out inconsistencies between specialties and drag them out in the open for everyone to see.

This might help but not for long. Sooner or later these 'generalists' will wind up being just as position-proud, hidebound and arrogant as the most straitlaced of the existing crop. It is merely another example of mankind's macho determination to dominate all other humans.

But it might help for a while.

Meanwhile, I invite you to consider where to go from here.

Think about it a bit and the answer becomes obvious. There cannot be many such stars. A few thousand thinly strewn among the billions of halo stars sounds about right.

Rather surprisingly, precisely this sort of stars has long been known. The first conference on faint blue stars was held in Strasbourg, France in August of 1964! The conference was an outgrowth of the 200 inch Palomar photo survey which took place in the later 1950's. By now perhaps 300 have been catalogued. There are a few which are uncertain, but there are also vast reaches of galactic halo space which have not been systematically examined so it would be unwise to place too much credence on any numbers cited here.

These blue stars, called Humason-Zwicky stars after their discoverers, were thinly scattered toward the galactic poles (i.e. far out in the population 2 region). They are irregularly variable and have been known to flare brightly for a day or two before dwindling for weeks at a time. During their flare period they are known to exhibit cooling down into the yellows but as they fade they recover their more usual blue, or even ultraviolet. In short, they fit in squarely with my diagnosis.

Paradoxically this bothers me. I have been aware of the faint blue star conundrum since the University of Minnesota published the proceedings in 1965. The concurrence between my conclusions and the 'Blue star' data make me unsure whether I may have skewed my logic to conform to the actual data or simply have reasoned the matter out independently. Such close conformity is not normal so I urge anyone who reads this to treat my last few paragraphs with caution.

Assume I am correct. If so, these faint blue stars would be a transitional phase where the core region of an erstwhile black hole is still laden with photo plasma but the translation of the Cantorian straight line photos has been reduced to a trickle while the surface of the hole is progressively expanding and in the process is bleeding off heat *en route* to graduating into an ordinary neutron star.

Am I correct in this analysis of black holes?

Probably not!

Astronomers know it is virtually axiomatic for the initial interpretation of any phenomenon to be wrong.

There are simply too many unknowns out there in the universe, and just about everyone wants to get in on the act. One solution after another is advanced and later discarded. It may take a few centuries but ultimately the reality becomes accepted... At least occasionally the reality is accepted.

I have broken new ground here. I have explored concepts and traveled highways and byways never before trodden. It would be a genuine miracle if everything I have ventured here was to be proved correct, but I would like to believe I have moved to mend a few cracks in our comprehension of reality.

CHAPTER II

WHITE CLOUDS

What would we call the diametrical opposite of a black hole? Could it be a white cloud? This might be an appropriate title for our next excursion into galactic cosmology.

But why a 'white cloud'?

Our saga begins in the earliest days of human record keeping; a time antedating the Sumerians and the pyramid builders. Gods of all sorts and sizes bestrewed the world like a plague of locusts and nothing ever happened by accident. Today we have largely forgotten that one of the Christian devils was named 'Baalzebub', which translates literally as 'God of flying pests, while 'Lucifer' translates as the 'God of Light'! Oddly, the Greek counterpart to Lucifer is Prometheus, the god of fire, who was bound in chains atop a rocky peak for his sin in taking pity on the misery of mankind and giving us the gift of fire. And these were but two examples where the names persist long after the origins are forgotten.

There were the gods of the heavens, gods of the underworld and a myriad other gods stalking the land or swimming the seas. In heaven the Sun was the big god, but he was matched by the lunar goddesses, 'Selene', 'Isis', 'Hera', or 'Astarte', depending on your nationality: and in those days any city boasting a population of two thousand or so called itself a nation and revered its own family of gods and godlets.

Hordes of greedy priests busied themselves with the task of learning the will of the gods and in the process benefitted themselves enormously, competing with the priests of other gods in wealth, pomposity and authority.

And how did they do it? It was through the simplest way imaginable; by interpreting omens and portents. There they were supported by the all too human refusal to accept blame when things go wrong. There always has to be someone or something else to blame. If you doubt me just peer beneath the surface of any modern general or bureaucrat or teacher or politician or lawyer. Always and forever there is the piteous plaint, "It's not my fault. He didn't do what I wanted him to do so I had to punish him!."

When it came to pronouncing omens or portents or divine decrees these humble servants of the gods were aided by the study of heavenly signs and the science of astrology was born.

It quickly achieved precedence over the other sciences or omens. After all, the big gods lived among the heavens. Littler gods roamed the Earth and seas while the dismal gods lived beneath the Earth in a state of perpetual deathly gloom.

We dare not underestimate the power and authority of omens and portents. To this day more people worldwide believe implicitly in astrology than disbelieve in it. It has endured since long before the written word even existed and bids fair to continue for many additional centuries in human belief patterns until our final extinction as a species.

The theological implications of

ancient astrology have bedeviled astronomers well into the modern era; lurking as a subconscious bias deeply ingrained in the most primitive regions of the brain... astronomer's brains among them.

What am I talking about? I am not speaking of drawing up charts and signs and pretending to know what the gods may have in store for mankind. Any astronomer would recoil in exasperation at the idea of charting human characteristics or fate based on the natal position of Saturn or the phase of the moon, but there is still a tiny hint of the ancient faith resident inside his skull.

Consider the reality. According to faith, a god must be perfect in all respects. This is item number one in the litany. Given this perfection it must follow that the god is incapable of manufacturing imperfect products.

An oval is an imperfect circle. The Copernican astronomy was initially rejected because astronomers and churchmen objected to the idea that God might not have been employing perfect circles for orbits. His universe cannot be a messy affair. Everything must have a precise origin lacking all hint of contamination. 'From the clock we deduce the existence of the clockmaker'. In short, perfection breeds perfection.

A high ranking prelate refused to accept Galileo's invitation to peer at the sun through a telescope to prove his contention that there were sun spots, uttering a contemptuous rejection that "He would rather believe that a telescope lied than accept the idea that God's perfect creation should have measles".

Nor should we neglect to mention that Rutherford's initial depiction of the atom required a nucleus (translate this into 'sun') surrounded by satellite electrons (planets) all traveling in perfect orbits as proof of their divine origin.

Now translate such a bias into astrophysical terms.

Everything is trudging along quite nicely when we start talking about the pathway to becoming a nova. After all, didn't the whole universe start with a bang? So a series of progressively littler bangs hangs right in there. God's work is not only perfect; it is also an inspiration and a warning to his human pets. In other words, don't trifle with perfection. Don't even imagine it!

Suppose we accept this thesis and see how it plays out.

The bias is subtle. Most would deny that it even exists, but consider the process of manufacturing stars. Put an enormous amount of hydrogen gas together in one spot and the resulting gravitational pressure can only be countered by opposing heat energy. The solitary source of so much energy is nuclear fusion. Now you have a star. The more gas you have the hotter the star is and the faster it must burn in order to push back against the accumulated gravitational force of the gas. With only a few little kinks in the nomenclature, such as Type B stars, we arrive a neatly descending scale where the larger the star the greater its brilliance, ranging from the largest and hottest down to the smallest and coolest.

It is the epitome of perfection; an

orderly descent from the largest to the smallest. Any theologian would gloat with smug self-satisfaction at the arrangement. It is further proof that God only works with perfection. Additionally, the astrologer concealed behind a few astrophysical minds would rejoice at having proved once again the ancient truth which has been so often been dismissed by pagan vulgarians.

The trouble is, there is a logical disconnect here. For the first generation of stars it is fairly evident that the gas accumulation bit is a necessary preliminary. But this says nothing about the second and subsequent generations of stars. Nor is there any *a priori* reason to postulate that later generation stars must follow the theological bias. In fact, there is every reason to dismiss it as a caveman era myth!

The interior of every star is exceedingly complex and my goal here is not to go into the complexities of nuclear fusion. Many texts, dating back to Chandrasekhar in the 1940's have amply resolved these issues so there is no reason to retrace their tracks. Suffice it to say that the core of every star conceals a fusion reactor which is busily synthesizing new elements in order to extract the needed quantities of energy wherewith to resist the gravitational pressure of the atmospheric envelope blanketing it.

Once the core reactor exhausts it's supply of fusionable small atoms and collapses, the temporary energy generated by the frictional forces not only expels 80 to 90 percent of the blanket gases but also produces a small number of transferric nuclei, running all the way up to californium. to serve as higher elements once matters settle down and the free nuclei, otherwise known as multiply ionized elements, are part of the ejecta.

We might easily stop here, but as the TV huckster screams at you in hope of peddling his nostrum, "But wait, there is even more! In addition to what we have promised we will even double your order of 'Stomach Gunk and Hair Restorer' at no added cost to you!"

When a star goes nova it is unrealistic to decide that there is some sort of cosmic filter which prevents all of these heavier atomic nuclei from being blown away in the explosion. Were there really such traps we would inevitably wind up with a few billion cast iron erstwhile stars floating around in space with only a scattering of still radiating stars to illuminate them!

Of course the whole idea is an absurdity. While it is remotely conceivable that a smattering of such stars exist in odd corners of the universe they are most certainly not common.

In anticipation of later discussion I think it highly possible that the innermost cores of fourth and later generation stars such as Earth's sun will prove to possess exceedingly dense nuclei consisting of multiply ionized iron plasmas capable of modifying the production of some classes of neutrinos. But I am not convinced of my logic here so I content myself with the suggestion that it may be correct but not necessarily so.

What is certain is that significant quantities of heavier and ordinarily non-gaseous elements are among the ejecta driven out into space in novae.

There is a further little item to throw in here. Heat is a product of motion. The faster the motion of an object the greater its energy. Heat is energy. It need not always seem hot to our human touch, i.e., when expressed as vibrations attendant in the interaction between molecules in compounds or lattices in rigid structures.

A noteworthy consequence of this in stars is the fact that the surface of a star usually runs a temperature around $6,500^0$ while a few kilometers above the surface the temperature will typically run to 1,500,000 or so degrees!

This is merely a twisted permutation of the automobile tire paradox. When filled and sitting as a spare in the trunk, the air inside the tire is at the same temperature as the ambient atmosphere so the energy levels are in balance and there is no difference between the two.

Now mount the tire on a wheel and drive off. The outside air molecules in immediate contact with the tire may experience a brief elevation in temperature consequent to the agitation as they were jostled out of the way, but there is so much air out there that the heat generated is soon dissipated.

Within the tire the situation is different. Every molecule of air in the tire is continuously agitated, which translates into 'heat' in the popular jargon. In moderate weather this heat builds up until it arrives at an equilibrium with the confining rubber. If this equilibrium is not reached the air inside the tire gets hotter and hotter until the containment rubber is melted and a blowout ensues.

At the other extreme, any air valved out after an extended session on the highway emerges as a downright cool wind after shedding much of its kinetic heat while escaping the tire.

So how is it that the ejecta from a star gets hotter when it escapes confinement?

The answer is straightforward. It is a variation on the old 'desert sand' phenomenon. When the solar rays strike the sand it grows hotter. As more and more heat is pumped into it the temperature steadily rises. But this is not an open-ended increase. When a maximum heat retention capacity for the receiving material is reached it must reradiate all excess or it will melt down. Until that critical point is reached the receiving mass simply gets hotter and hotter. The upper solar atmosphere is stabilized around that temperature. For the most part this is due to the kinetic energy of the escaping radiation as the fleeing particles jostle and collide with one another, so there is nothing surprising about the difference in temperature.

Given a 1.5 mega-degree environment we find we are faced with an electromagnetic plasma and an accompanying process of spallation. This leads to a synthesis of higher level nuclei which can escape along with residual hydrogen. Once escaped from the solar atmosphere these ionized nuclei pick up stray electrons which we then allude to as oxygen, nitrogen, carbon, aluminum, etc. Included in this mix are a noteworthy production of water and simpler carbon compounds.

This is the easy part of the puzzle. Now matters start getting complicated. So suppose we start with the simplest problem and see where it leads us.

The distribution between Type 1 and Type 2 stars is notably different. As a rather imprecise rule, Type 1 stars typify the spiral galactic structure while Type 2 stars typify the globular, or halo part of our galaxy. Barnard's Star, which is only a few light years distant from our sun and thus well within the spiral arm region, is demonstrably one of the Type 2 class. However it is also in a well-defined, highly inclined and pronounced eccentric orbit which swings it far out into the Type 2 region most of the time, so it is not truly an exception to the rule.

This brings us to the crux of the problem. We clearly have two distinct systems in operation here, with the halo stars being generally the elder and more primitive of the two.

Having said this, I am now constrained to seek out which of the two alternatives to my conclusion applies. Spelled out, the first alternative argues that there has been a misinterpretation of our sense of direction of the spiral arms and they are actually spiraling into the heart of our galaxy rather than streaming out away from it!

Try this one on for size and see what ensues.

First problem: If the spiral arm material originates outside the galaxy and is being drawn from there into the core, where do the required iron, nickel and other heavier elements typical of spirals come from? Are they lurking in the dark between the galaxies awaiting a wake-up draft from the core of the galaxy to make their presence known?

A priori, this seems highly unlikely, but over the years any number of suggestions which have been rejected as 'highly unlikely' have come back to haunt us and made their protesters seem like academic fuddy-duddies. So we try another line.

If the spiral materials are indeed falling into the galaxy and bringing relatively massive quantities of iron sufficient to manufacture a few tens of million population 1 stars we would expect to see evidence of this infall coming from all regions and the perimeters of the globular segment of the galaxy would reveal at least a scattering of population 1 stars. Since we have found no such indications we must conclude that the idea of an infalling spiral is fatally flawed. Therefore, coming or going, it is clear that conditions at the core of the galaxy are critical to the development of population 1 stars and any associated elements common to our own solar system. Most or all of the heavier elements found anywhere in the galaxy are products of the identical conditions as those which manufactured spiral arms.

Now we ask what these conditions were, how long they have

persisted and what they are comprised of?

An inescapable implication here argues that our sun and our solar system were born and raised in a spiral excrescence some 5 or 6 billion years ago. Thus we now have one part of the problem more or less solved. But it is the easy part.

To investigate a few of the deeper problems it is necessary to return to logical fundamentals.

CHAPTER III

DYNAMICS

Take one problem at a time and do not try to avoid logic by pretending it is unnecessary. From the onset of my little excursion into cosmology I have been committed to the employment of logic rather than mysticism. Here is no exception. We have a situation where material is streaming out from the core of our galaxy, ostensibly from a 'black hole' from which nothing can escape.

This argument was addressed and mainly refuted in the opening chapter of Book II, where unwaved photos (dark matter) necessarily escaped since they were independent of our little universe and were inhabitants of the underlying mega-universe.

Spallation effects consequent to the environment around a hole seated in the midst of a surrounding cluster consisting of a few millions of stars regenerated the long discredited Olber's paradox to manufacture a region where spallation of the lighter elements would occur.

But now comes the problem of heavier elements and the manufacture of spiral arms to propel them away from the core region. This bids fair to be much more arcane than anything I have discussed thus far, so please bear with me if I ramble a bit more than usual along the way. Accurate analysis of spiral arms in galaxies is surprisingly scanty and what little there is largely confined to photographic nomenclature aimed at pointing to their physical appearances. This means we are starting with what is virtually a blank slate and there are problems all over the landscape. A few examples introduce us.

First problem: How do we produce elements ranging down from californium, to uranium, thorium, and lead, all the way to iron from ordinary spallation?

The problem seemed intractable until fairly recently, when I acquired an electromagnetically shrunken U.S. Quarter produced and copyrighted by Bert Hickman in 2007. I can do no better than to excerpt a brief paragraph by Edmund Scientific's describing the process.

"The Quarter was created using a technique called *'electro-magnetic forming'*. This interesting little technique uses powerful electromagnetic forces to 'shrink' a quarter to a diameter smaller than a dime. This process uses an invisible, but extremely powerful, pulsed magnetic field that evenly 'hammers' the coin inward with powerful shock wave, forcing it to shrink with a blink of an eye."

The process requires forces running upwards of 100,000 amps, providing some 5,000 watt/sec with the peak current within the coin approximating 600,000 amps. If subjected to even higher amperages further shrinkage is possible.

By direct implication electromagnetic forming can produce a myriad surprising effects not all of which can be anticipated in primitive textbook theories.

Now we return to the environment at the core of the galaxy

and consider the magnetohydrodynamics involved.

Start with ten million stars packed into a sphere perhaps 50 light years in diameter. As described in the last chapter, these stars are necessarily seeking some way to expand and contract simultaneously. In effect they are jam-packing the ambient space with pulsed radiation. But rather the 600,000 amps cited in Hickman's Quarters the pulsed radiation here might easily exceed 60,000,000 amps!

This would be more than adequate to produce transferric elements on a wholesale basis

If my analysis is correct then the problem is resolved. The presence of a hole surrounded by a family of core stars whirling in orbits around it not only creates a spallation zone in near orbit around the hole but induces a titanic magnetic flux which permeates the core region with pulsed radiation sufficient to manufacture nuclei running up to and beyond uranium.

Escaping from the core these nuclei accumulate stray electrons thereby deionizing themselves to create the heavy elements we find on Earth. Two problems solved; one to go.

The one remaining problem bids fair to be more intractable.
Why do we get spiral arms?
Part of the answer is obvious.
The human tendency is to perceive data as static. The stars at the core are imagined as simply hanging there in immovable majesty. When we stop to think about it we instantly recognize the utter impossibility of our idea. Stick a pin in our star to hold it in place and it must fall into the hole. It has become a 'moon'. And like any moon it can only keep its distance by orbiting the primary, ergo, the core stars must be in orbit about the hole; and they must be in fast moving orbits.

At some 30,000 light years from the core our sun is loping along lazily at 150 km/sec. Were it only 10,000 l.y. from the heart of the core it would be galloping around at about 1,350 km/sec! While not in the same league as light it is still rather fast.

In brief, there must be a pronounced equatorial bulge girdling the core and material being ejected from the core would be flowing along the track of the equatorial bulge. So there is no esoteric mystery why expelled matter should be streaming out from the waist of the spinning core. Neither is there any mystery in the curved track of the ejected matter as it races through the galaxy.

The material was ejected from the core region with an initial velocity sufficient to drive it to the perimeter of the galaxy. Some fraction was probably ejected from the galaxy altogether, in the process becoming roughly the equivalent of an Oort cloud drifting outside the formal boundaries of the galaxy. But the outbound material is continually slowed by the gravitation from the core as well as minor impedances along the way.

This says nothing about the lateral motion imparted by the core rotation. While there will be a gradual slowing caused by the inevitable collisions with slower moving material in normal orbits around the galaxy, these will be less frequent and require more time to drop into parity with the ambient matter.

So far, this is all to the good. But the central question remains untouched, i.e., why do spiral arms tend to emerge in opposing pairs? We might expect them to flare outward from the equatorial core in all directions and more or less simultaneously, but they obviously do not.

This is our mystery, and it would be interesting if I could offer a solution. Unfortunately, I cannot.

The problem is, there are simply too many possible avenues available. Spiral galaxies abound in our universe. Some almost seem to have sprouted like Athena, full blown from the brow of Zeus. There are ring spirals, twisted spirals, barred spirals, open spirals, aborted spirals, all strewn randomly throughout our little universe. I am compelled to regard the problem as indeterminate pending further data. Perhaps it is merely a flip of the coin type of problem but we must assume that differing events developing within the individual galaxies are responsible for each variety.

But perhaps the question is unnecessary. We possess enough data to allow a reasonably accurate reconstruction of the events leading up to the development not only of Type 1 stars but also of conditions within the spiral arms.

Take it from the top and do not fall into the error of thinking of the newly formed universe as some sort of a horn-shaped affair rambling along through the ylem. Instead you should think of an ordinary cardboard box filled with smoke. Punch a small hole in the side of the box, then tap it gently on the opposite side.

A small burst of smoke emerges and rapidly forms itself into a smoke ring. Then note that any residual smoke which seeps out afterwards simply curls off and dissipates. The smoke ring is therefore not tied to its point of origin by some sort of umbilical. The same would apply to our universe. In short, once the ring is formed the immediate connection with the ylem is severed. The ylem remains tied to a Cantorian lineal C set while the newly formed universe is enmeshed and operating in the Cantorian curved C set.

In effect, this means that any thought of backtracking to a point of origin is not only hopeless but also rather absurd. No 'Big Bang' is possible.

As a final item, please bear in mind the muddiness of the universe. There is no hint of solemn, stately Wagnerian order to the affair. It is incredibly violent from start to finish.

Once these two delusions are eliminated we are free to complete our progression.

First comes the puff of smoke generated out of the mass of ylem accumulated in one region consequent to Kolmolgorov turbulence formulas.

Now ask what the conditions must have been within the still evolving torus?

There are a few pertinent analogies right here on our tiny little planet. We have tornado weather where as many as seven or eight tornados may be seen by a single observer in a single instant. We have hurricanes. Most of us have seen and heard bolts of lightning in the heavens. Sheet lightning is commonplace in desert or prairie lands. Ball lightning, otherwise known as St. Elmo's fire, has been known for centuries.

Before any objections are made, yes, there are instances where local conditions and solid masses are contributory factors in generating some of these phenomena. But this is not a fatal objection. Conditions within the torus are a million times as wild as they are anywhere on Earth.

Consider the phote. Each one possesses a minuscule mass/gravity. Each also is characterized by a weak polarity. At the beginning that is all there is. But now things are changing. The photes are being warped and contorted by strange patterns which combine to stress them in every direction. Chains of photes briefly link together but a moment later they are violently torn apart. A few tens of trillions are torqued beyond recognition as the head of the chain is twisted back upon itself then twisted like a tiny rubber band which has been contorted into a knot which we have named a 'proton', or perhaps merely an electron if the torqued chain is shorter by a few hundred photes.

Throughout the whole of the emergent universe visions of light and color are on every hand if there were there eyes to see them. After all, these photes are also light, and under the conditions outlined here they should be running all up and down the spectrum.

The vista can best be perceived by first visualizing yourselves in a glass space vehicle and striving to comprehend what we are seeing.

Before you: To either side of you: Behind you. A globe of lacework filaments sleets iridescently from one horizon to the other, gleaming with every color of the spectrum. There, off to one side are muted purples and maroons concealing the nesting place of a star just now birthing itself. Over there is a brilliant azure blue behind which lurks a newly flaming giant star. A delicate coral pink halos a nearly crimson red where the cloud conceals an unstable star which one day soon will erupt into a titanic supernova. Pale greens and blues and ivories stream across this newly hatched universe like unto a terrestrial aurora, only here they are arching from horizon to horizon like vaults of a celestial cathedral.

Between the luminous filaments separating one from the other, trail lanes of chocolate browns and blacks, along with dozens of hues grading off to either side. Here and there a nearer dark cloud crosses an intertwined hatch-work of lustrous bands shaped much like thunderheads against the background of our terrestrial sky while elsewhere brilliantly noctilucent clouds stand in stark contrast to the blacks and browns behind them.

Titanic bolts of lightning powerful enough to devour entire suns flash randomly about as they bridge distances hundreds of times greater than the distance of our Earth from the sun while appearing as mere sparks dappling the heavens from where you are standing. A hissing, crackling, spitting sound punctuated by muted roars of thunder, mysterious distant explosions and dull, rhythmic thump-thumpings are relayed to you from the sensors of your ship as its computers strain to decipher the myriad radiations being received in the only way they can.

No artist could hope to capture the chaotic magnificence of the spectacle. Perhaps the nearest approximation is El Greco's justly famed "Toledo". But even the best will fall short of the reality before him.

I have not been exaggerating in my description here. In sober reality I too have fallen short. This was at the beginning of our universe. All of the millions of galaxies and untold trillions of stars were jam-packed and crammed within a volume of space probably about ten million light years across… which is mere spitting distance as our present universe goes.

By now just about everyone is pretty confused. What has this depiction of the birth of our universe to do with the emergence of spiral arms in galaxies?

But consider our little galaxy. Everything I have described in these few paragraphs is replicated in the galactic core. The same chaotic turbulence is echoed on a smaller scale. Strangely, it also appears here in our own solar system, and even on our own home planet! Are we not arrogantly presumptuous when we scoff at the idea that the chaos attending the birth of our universe should be tamer than what we see all about us in our own tiny niche of the single, medium-sized galaxy?

This is the point in time when the conditions which led to the appearance of our galaxy and subsequently to the creation of the galactic spirals were generated.

But our tale is not yet told. There are some interesting avenues yet to be considered. I have spoken at some length of the conditions at the center of our galaxy and have made some crude, off the cup estimates of its size and configuration. The spiral arms may allow us to refine these estimates, and in the process may help us understand why there are only two spiral arms rather than four or eight.

The dimensions of the conventional black hole, as derived from minimalistic calculations, are not cast in iron. They merely point to the least mass which can produce a hole. In no way do they address dimensions where two, three or a dozen holes are blended into a single hyper-hole. This might occur if two holes in close orbit about one another ultimately spiraled together, perhaps setting the stage for additional collisions down the line?

If we postulate that a routine, minimal hole has a diameter of 4 km and it merges with another hole of like diameter can we then argue that the resulting double-sized hole is further compressed and therefore is only 3 km in diameter, or do we take the opposite tack and posit that it is now 5 km in diameter?

Given the ylem/phote model posited here the interior of a black hole must be crammed with tightly packed photes arrayed in six Cantorian curved infinity geometries plus those obeying an added straight line geometrical infinity and thus are dropping back into the primal ylem and streaming outward as dark matter.

By implication, the hole cannot get much smaller than the minimalist limit or it will expand and lose its black hole status. But this says nothing about the possibility of hole growth by the acquisition of new matter. Should this occur we must expect the escape rate of straight-line photos to increase dramatically, but not necessarily at the same pace as the acquisition of new mass, in which event the hole diameter will be increasing.

Now pause for some more time-line calculations. We have reliable data establishing our solar system somewhere before 4.5 billion years ago. What the few observations which focus on the spiral arms would suggest is that there has been no significant fluctuation or interruption in the rate of ejection from the core at least during that interval. Neither do we see any indication of earlier major fluctuations.

Throw in a few more calculations and see what we find. Our solar system is approximately 30,000 light years from the heart of the galaxy. But it is also about 20,000 light years from the outer fringes of the galaxy.

Metal rich population 1 stars are seen all the way out to the perimeter, ergo there is ample reason to believe that the spiral arms have been in existence for a minimum of around 9 billion years, during which period the hole has spewed out roughly 50 billion population 1 stars. Rounded off, this means the hole must be spitting off mass at the rate of 5.6 full size stars every year of these 9 billion years!

I may be wrong here, but I am treading on virgin territory which lacks any established landmarks so if I am off by a few orders of magnitude I hope I may be forgiven my inability to insert the required decimals. But no matter what, this is still a formidable number. Perhaps a touch of additional refinement will help resolve the problem.

Olber's paradox and its effect on perimeter stars around the core comes to mind. As pointed out earlier, these stars are in something of a quandary being compelled by competing forces to expand and contract simultaneously. In considering this I was reluctant to attempt a solution and thus was impelled to opt for a minimalist solution. But here it is clear that anything which is rational rather than pure mysticism and offers help in resolving the problem must be pursued.

Olber is not likely to provide a complete solution but it is evident the whole envelope surrounding a hole at the core must be a pea soup of conflicting electromagnetic forces. Spallation effects must be multiplied by as much as 10^{20} or even more. In the process many, if not most, of the escaping lineally oriented 'black matter' photes are contorted and thus drafted out of the Cantorian lineal C-set and into his curved C-set. This increases the spallation effect.

But here again we must come to grips with the human inability to think in terms of big numbers. To virtually all of us 10^{27} is mentally indistinguishable from 10^{26}, and even if we choose to write it out a 'one' followed by 26 zeros is only marginally less than a 'one' followed by 27 zeros. Comprehension can only exist in the select few individuals who set out to become intellectual robots.

So how do we get around this dilemma?

An old Islamic saying comes to mind. "If the mountain will not come to Mohammad then Mohammad must go to the mountain." Rather than looking to the large we must look to the small. The phote is the smallest object in the megaverse. Shrink our bodies and minds until we can book passage atop a phote and experience what it must experience if it is lodged within a mammoth black hole!

Before we set off on our journey it is only fair to alert all of our readers that the writing will be both murky and confusing. This is inevitable. The micro-universe we are entering is the very essence of chaos in its irrationality. Regardless of what bean counters may pretend, you cannot program randomness any more than you can convert pi from an irrational number into a rational one. The best we can do is pluck a phote somewhere in the heart of a hole, outline its adventures and hope for the best.

We must start by disabusing ourselves of any notion that the interior of a black hole is dark and dismal! It doesn't work that way. A black hole is only black from the outside. Inside it is blindingly white. Every ray of light which strikes it is instantly absorbed and converted into heat and brilliance. Temperatures may rise into the millions of degrees centigrade, and with no way to escape their prison it can only keep rising as the stream of light from the ambient stars continues to pour in. Like it or lump it, this is the environment in the interior or a black hole, and equally to the point. There is no conceivable way to destroy a phote since it is energy per se. You cannot release energy by destroying it!

Now we are able to hoist our anchor and set sail on our little journey through the hole.

Though no doubt incorrect we assume the uncurved phote is somewhere in the vicinity of 10^{-9}cm in length and has an unstressed diameter

77

roughly a third of the length. This is not a purely arbitrary number since it is close enough to the value of the gravitational constant to suggest a connection but large enough to reduce the number of zeros to a manageable length. 0.00000001 is much easier to put to the pen than an absurd .000000000000000000000001, and I am lazy enough not to appreciate the effort of playing with that many zeros. It seems almost congressional in its stupidity and I am horrified at the prospect of being lumped together with those chaps. Besides, doing it this way I am indulging in the time-honored tactic of covering my posterior in the event some unkindly soul wants to fault me for making some blatantly flawed statement.

In theory we are traveling at a speed of perhaps 300,000 km/sec. But this is more a theory than a fact. Less than a billionth of a second into our journey our phote rams into another phote galloping along on its own track. It promptly recoils into yet another phote, where it is battered anew by a third phote, followed by a myriad other collisions, all of them coming from a myriad directions and configured to the requirements of 6 curved dimensions plus the straight line dimension.

Hour after hour, day after day the bombardment continues. After the first full day on our journey we have strayed perhaps a centimeter from our starting point!

Hour after hour, day after day the relentless pummeling persists, with no let-up in the bombardment. Nor is this the ending of our little tale. With every collision there is an inevitable deformation in the shape of our vessel. When we boarded it our ship may have been configured as an inhabitant of this universe, but the first collision may have torqued it into an antimatter object characteristic of our opposing universe while the next after that may have configured it to be an inhabitant of one or another of the lateral universes.

Somewhere in the mix our photo momentarily acquired a few hangers-on and rolled itself into a tight wad, there to proclaim itself an electron and thus the lord high poo-bah of the place.

This lasted maybe another millionth of a second before a thoughtless phote put an end to the nonsense by shattering it into its component parts and leaving it back on square one.

Over and over the cycle of random collisions and fleeting linkages repeats itself. How long does it last? No theory exists to measure it, but analogous time frames suggest that the escape of a single elementary particle traveling from the core of our sun to the surface may take as long as 2,000 years! Using this as a criterion after discounting liberally for the differences in size and length of the journey, the estimated time required for a single phote to be transformed into a lineal C-set where it promptly merges back into the primal ylem of the megaverse comes in somewhere under a decade

But this does not necessarily mark the end of the line. More accurately, it marks the end of the prelude. While it is possible our phote may continue unhampered and revert to being merely another component of the ylem, this is highly unlikely.

The instant it emerges from the crust of the hole it is assailed by the infalling light from the outside galaxy. Where the photes within the hole are of 7 different geometries, those falling into it are either left-oriented or straight line photes with no place for the other 5 configurations. Any strays which might sneak out for a few moments of freedom from the hole are faced with immediate decay, in the process contributing immensely to the satisfaction and egos of the bean counters who have almost nothing better to do with their time than play around with their giant accelerators and feel important.

The straight line photes emerging from the hole are reduced to one of three options, i.e., the first being to evade any co-option and remain in in the ylem, albeit not part of this universe, just as the air through which a smoke ring passes is not part of the ring.

The second option has it linking with other straight-line photes where it exhibits the qualities of magnetism and gravity and becomes part of the giant magnetodynamic plasma field which engulfs the hole.

The remaining option is perhaps the most interesting to us. Here the newly liberated phote encounters a growing linkage of left-curved photes and is contorted by the ambient plasma within the spallation region surrounding the hole.

In short order the assemblage emerges as a brand new electron... or perhaps even a proton. Where such linkages would be torn apart in short order if they occurred inside the hole, this is not a necessary fate in the spallation zone. Here the flux density is four or five orders of magnitude less than within the hole and the disruptive effects of the alien geometries are absent while the kinetic temperatures which had been dramatically reduced by the expansion of the playing field attending the escape from the extreme compression regions within the hole are starting to rebuild.

All these things are occurring in chaotic profusion within a region stretching over a volume probably building up from a starting point roughly one kilometer from the surface of the hole and extending outward to a point roughly 500 to 700 km distant from the hole where the temperature has decreased to a point below criticality while the magnetohydrodynamic plasma density has continued to increase and a second creation process appears on the scene.

    Before addressing this it is needful to point out that in the 1970's or 80's spectral lines unique to the element Californium were detected in the expansion sphere of a distant supernova. Since elemental Californium is a transuranic element whose life expectancy is measured in seconds it is evident that structural build-up of heavier elements can continue well beyond what we would anticipate from a spallation region, hence the inferred existence of the third region extending beyond the traditional spallation zone surrounding ordinary Population 1, type O stars.

I confess I am not entirely happy with the Californium data. There are too few confirmed tidbits of information available. For example, what was its actual distance from the collapsing core at the moment of discovery? Was it a one-time event caused by a collision with a stray bit of cosmic radiation which just happened to be in the neighborhood? Or more remotely, some work on the stability of transuranic nuclei in the 1930's and 40's pointed to an 'island of stability' around a hypothetical element number 126, which was supposed to mark another boundary for a region akin to the rare earth series amidst the lighter elements?

I recall filing this datum somewhere in the recesses of my memory where it remained isolated for perhaps 70 years before surfacing here.

Since I cannot vouch for accuracy I only mention it because it ties in so well with the idea of a necessary second creation zone formed outside further out from the core hole than current astrophysics suggests.

Rather than focusing one temperature based spallation which creates protons, neutrons, electrons and lighter elements, this region is characterized by extreme electromagnetic fields capable of generating the heavier elements up to and including the transuranics.

These fields are perhaps more appropriately addressed in the next chapter, when the focus centers on the spirals per se.

A sense of honesty prompts me to admit that this chapter has been without doubt the most difficult and frustrating one I have ever written. As I pointed out at the start, it was murky and disorganized.

True chaos cannot be organized and how can anyone hope to convey a sense of reality to the presence of untold trillions of photes jam packed within the overwhelming luminosity of a black hole, jostling each other trillions of times a second and being deformed haphazardly into one of seven possible configurations during each collision? Next add to the chaos by pointing to the prospect of one-seventh of these photos momentarily dropping out of the hole while still inside it, with most of these undergoing a fresh deformation and being back within the hole while those which pass beyond the spatial bounds of the hole are buffeted by a fresh assault in the spallation region?

How can anyone describe it? I have done my best. Perhaps another can do better.

CHAPTER IV

WHITE LANCES

I suspect that the birth of opposing spiral arms is rather prosaic in origin and is actually a naturally balancing mechanism,
but this is easier to say than to prove. Why not 3 arms, or 4, 6, 8, or even 12 radials?

Actual observations of our own galactic arms support a conclusion that a dozen or more partial spiral arms have appeared over the ages or are even now forming. Unfortunately, the verdict is not yet clear and other more distant spiral galaxies do not invariably support such a hypothesis

For that matter, according to conventional theory holes are supposed to suck matter into themselves, not spew it out; especially in such copious amounts in a continuous stream over a period of at least 9 billion years. So how are we to account for that?

Much of this was discussed in the earlier chapters, but our mentioning here is aimed more at reinforcing the complexity of the situation by pointing out how the solution to one problem can lead to a new quandary where escaping one trap merely pushes us into a new trap.

The only way to resolve these questions is to take nothing for granted. Start at the beginning and work forward from there. And the only available starting point is the hole itself. It must be spinning on its axis; and spinning exceedingly rapidly at that!

Consider the logic involved. Here we should bear in mind our conclusion that every phote must possess a minuscule trace of mass/gravity as well as a polarity. Possessing mass it must be subject to the laws of kinetics and therefore will travel at a speed somewhat less than infinite. Possibly the difference amounts to as little as a single centimeter per second less than C, in which event we may infer what C may be, but we will never be able to determine it directly since it is inherently impossible to measure or assign any limits to an infinity.

With this caveat in mind we may move on to a critical factor in our reasoning.

Consider the elementary equation $O=gr^2$. I am deliberately avoiding conventional notation in the interests of keeping it simple. Translated, the orbit, 'O' of a body around a primary is equal to the square root of the gravity applicable at the distance by the distance. This was discussed in book 1 but it bears repeating here. Now let us see how it applies within a black hole.

If the speed of light Is truly equal to C then the equation for a beam of light orbiting a hole at the precise surface where it is absorbed by hole gravity must obey the equation $O=Cr^2$! In other words, at the speed of light Cantorian lineal infinity precisely equals his curved infinity and pi has become a rational number! The two conflicting C sets have merged.

Here we are faced with two alternatives, both of which can be sustained logically and both are equally attractive..

The first airily dismisses the whole affair with a wave of the paw and a summary statement that "of course this is true. It merely proves that ultimately every infinity merges to become a single infinity". The second is more prosaic and takes its cue from our conclusion that the phote (and thus the mass/gravity} cannot quite achieve C, in which event we might perhaps arrive at an equation where the operative number for the radius is 298,000 km minus 1 centimeter, in which event the problem is solved without recourse to off-the-cuff assertions.

In keeping with my avowed determination to construct a workable cosmology based on logic rather than theology I have opted for the second alternative. If this is too uncomfortable to be endured I am sorry and can only suggest you discard all I have said and come up with anything which suits your prejudices.

As something of an aside, I had no ulterior motive in selecting a one centimeter shortfall from infinity in the speed of light other than to illustrate how a tiny, virtually unmeasurable difference may lead to momentous consequences. The declared actual value for the speed of light has been "determined" so often and with so many variations that we are forced to conclude it is somewhere in the vicinity of 299,793 km/sec... plus or minus a thousand or so km/sec. You pay your money and take your pick. As long as you stay slightly above 297,000 km/sec you are probably in the right vicinity though 298,000 km/sec is most likely closer to the mark.

Now we play a few games involving numbers and concepts, without pretending to be accurate so far as the numbers are concerned. Our only concerns here are the implications resulting from the combination of the two elements.

Start with by postulating that the Cantorian 'C' set denoting infinite speed commences at precisely 300,000 km/sec while the speed of light is 298,000. Both of these assigned values may be sustainable but it is probable that one or both are incorrect. Neither is important here. Our concern is the implication of this postulation when applied to a hole.

I have already pointed out that the roiling rotation of any smoke ring universe not only imposes one and only one curved C set on any photes subject to the laws of toroidal geometry while the ylem atmosphere through which the torus is moving continues to adhere to the straight line C set and constitutes the dark matter of the universe.

The internal rotation of the torus implies a slow rotation of the gasses which are in the process of coalescing into stars. This must be true even if other forces are also at work.

But before a hole can be formed there must have been an enormous cloud of gas adrift in space. Somewhere in the midst of this cloud is a point where the pressures from all directions converge and the gas begins to clump in upon itself. A gravitational nexus is formed and we have the beginnings of a star. Ultimately the pressure of the infalling gasses reaches a point where the core region ignites and the star is born. This is the zero point not only of the star but also of our galaxy. It plays an integral role in the evolution of the smoke ring which constitutes our universe and similar galaxy-forming events are occurring everywhere within the ring.

So much for the easy part. Now we are confronted by a few trickier problems.

Since we have defined the speed of light as being less than the infinite speed of a Cantorian C set while the Schwarzschild singularity is a boundary where the mass/gravity properties of the hole engulf any light, mass or particles impinging on the hole's domain we find ourselves confronted by two, mutually exclusive alternatives. For one, we may assert that Schwarzschild stumbled and the whole premise of the black hole is fatally flawed. For the other we may assert that his premise was correct but incomplete.

Opting for the first alternative leaves us in a quandary. There is obviously something strange going on at the core of our galaxy. Variations of the same something at work in the cores of roughly seventy percent of the observed galaxies of our universe so it is by no means a rarity which can be shrugged off as a mere 'glitch in the work' of manufacturing a galaxy. So if a black hole is not the answer what else can it be?

By now it should be obvious that I am fumbling about. Since I am resolved not to dissemble or evade issues and to accept my failures I can only say that many questions regarding spiral arms seem intractable given the approach I have used. But this does not mean I have failed. It only means I must go back to square one and analyze the foundations of my reasoning. Where have I made a false-to-fact assumption? Is there a different approach I ought to have used?

On retracing my steps item by item it slowly dawned on me that I had made precisely the same error I had derided all the routine bean--counters for making. I had failed to start building my thoughts at the beginning. I had blindly accepted an omission Schwarzschild had made without first checking its validity.

It was literally a 'go back to square one' thing.

'Black holes' do not necessarily form from collapsed matter, nor are they necessarily manufactured by stellar behemoths. At least in fundamental theory they need be no larger than the (.) at the end of this sentence. They may even be no larger than the nucleus of a hydrogen atom! All we need do is postulate a gravitational or electromagnetic field capable of confining impinging outside entities so they cannot escape!

It is that simple. But it is not entirely that simple. There is more to be said

Start by playing an imaginary game with physics. I begin with the promise that there are no flying saucers, no matter transmitters without receivers at the other end and no magical incantations or mantras anywhere in the mix. Given this, let us see where it leads us.

Postulate an empty point in space. And when I say empty I mean empty. It is simply totally void.

Being void it is essentially dimensionless since there is nothing to measure it against so the point is equally dimensionless.

Atop this dimensionless point we place a notional mass of any strength we choose and assign it a notional gravity of the sort routinely posited by modern theological physics. Next throw in a speed of light notionally posited as equal to the Cantorian C; otherwise known as infinity.

In effect, we are dealing with a purely notional mass, an equally notional physics, and a probably notional speed.

Stir this mix well and see where it leads us.

First question. Have I any notion where all these notions are leading me?

The answer to this is 'I do'. So let us see just where I am being led.

First answer. The philosopher/theologian/mathematician Descartes sought diligently for some island of absolute certainty anywhere in the universe. At length he arrived at a solitary answer which has never been successfully refuted in the centuries which have elapsed since he advanced it. *Cogito ergo sum.* 'I think, therefore I am!' Some entity is doing some thinking.

Unfortunately, after arriving at this certainty Descartes lapsed into the realm of theology. Eschewing the solipsism of the politician or robber baron, Descartes denied that he was God and proceeded from there.

A God can only speak truth because whatever He says becomes real, i.e., God said 'Let there be light!' and light became real.

This is a nice thought, but to think implies a mind to generate the thought, which in turn implies a body to contain that brain, etc. In effect there must be an outside reality otherwise the whole thing becomes an absurdity. The nearest any theology has come to this is the Buddhist 'Maya' or delusion in which a soul is ensnared into believing it is encased in a body so therefore it is encased in a body. Possibly Zoroaster acquired this theological slant from the Mahayana model Buddhism, but it is also entirely possible he arrived at it on his own.

Either way the result is the same; there must be an outer reality in play.

The idea of a purely notional mass is therefore invalid. Call it by whatever name we wish there must be an outside reality, ergo we now are confronted with a real mass to which are appended a notional gravity and a notional speed.

But this does not mean we are prohibited from employing a notional mass so long as we accept the truth that in reality it is purely a logical ploy.

Now we restate our little equation and introduce an utterly dimensionless point into this empty space and then introduce an equally imaginary mass into that. Postulate that gravity is generated by this mass. As stated it is unimportant whether this gravity is merely a warpage of space, an imaginary stream of gravitons or a phote jet emitted by the mass. All we are interested in is creating a condition where the gravity emitted from the mass equals the speed of light.

Assume that the outer boundary of this condition is equal to one a.u., with the 'a.u.' being the mean distance of Earth's orbit about the sun.

As stated we have merely arrived at a notional black hole despite the fact that given the limits of our imaginary universe there is no conceivable way we could know this.

Our earlier analysis has proved that this presumption must lead to a paradoxical conclusion that pi is no longer irrational but simply merges with such other irrational numbers as the square root of 2, etc.

It would seem from this that we have reached an impasse, but closer examination reveals a paradox contained within the paradox.

We have a notional black hole whose throat is now fixed at 1 a.u. so we have a notional point of mass which acts once the origin of a black hole is fixed. But if the speed of light equals C then pi is necessarily rational, which cannot be in our universe.

Now take the next step and postulate that this condition is purely local and the laws of physics disappear within holes so the paradox is unreal in the normal universe. It sounds plausible, but is it?

There is a simple test which reveals the truth. It is easily overlooked because it is so obvious only a highly disciplined mind could ignore it!

Since the throat of the hole is one a.u. from the origin then any effort to apply the simple approach to the circumference of the hole must equal the value of C. This is easy. But now we pile on more mass to our point of origin, adding to it so the throat now appears at 2 a.u. Lo and behold, the same condition applies. The orbital speed implied here remains C!

Regardless of any effort to avoid the impasse the result must be the same. The only conclusion we may advance as an attempt to remove the objection might be to argue that our entire universe is a black hole and we are stuck within it, but closer inspection indicates that even this is implausible. For it to work we would be compelled to introduce the mathematical and technological apparatus of everyday life into the hole, which would inevitably lead to endless paradoxes all up and down the line; and which in itself would be tantamount to creating a Pandora's box of problems.

Anyone seeking to pursue this track has both my sympathy and my blessings, for whatever they may be worth. For my part I take it to imply additional logical proof that the speed of light does not quite equal C so all the paradoxes promptly vanish.

This approach has other benefits which are not obvious at first glance. But first we should dispose of a remotely plausible alternative which supposes that the speed of light actually exceeds C so light is transluminal.

In general I suspect this idea will be confined to compulsive nay-sayers whose chief aim in life is to refute any logical argument they may stumble across. As a rule their intelligence exceeds their common sense but on occasion they make cogent arguments so they merit our respect, if not adulation.

There also happen to be precedents for this idea. From a strictly scientific standpoint {which was pointed out in book I of this cosmology} the original presentation of the tachyon hypothesis was meant to 'balance' the equation of C by offering a descent mode where the other side of C reduced to zero. And what is this if not a transluminal event pointing to a something on the other side of C?

The other effort at evading the limits of C is less scientific albeit more interesting. Most of space science fiction seems to be premised on having space ships flittering around in the universe at multiples of the speed of light.

There is even a possibility which we acknowledged in Book I where the value of C may be different without interfering with the value within the toroidal universes so there is a bit of wriggle room here. But despite these potential conditions where C turns out to be slower than light they still cannot affect the reality that within our universe C must exceed the speed of light to some small degree.

On this note we finally have all our ducks in a row. The parameters are established and it remains to assemble them in order to obtain a clearer picture of the spiral structures and their place in galaxies.

Make no mistake. It will not be an easy task. In point of fact it will be an inordinately difficult challenge; one which we may not be up to facing successfully.

But we still have to try.

CHAPTER V

THE ASSEMBLING

Begin at the beginning.

Step 1 is the realization that the majority of all observed galaxies are spirals and most of those appeared in the earliest days of our universe. A caveat here argues that spiral arms tend to be more luminous and thus more easily visible at a distance than globular galaxies so the ratio between the two types may be skewed.

Step 2 notes that there must be at least two types of black holes. The most pronounced of these will be holes formed at the cores of these galaxies. A second, less pronounced source is one deriving from the collapse of super stars strewn about the usually type 2 metal poor stars typical of the first few generation galactic stars.

On analysis we find that core holes may continue to grow in mass and strength while perimeter holes are restricted in their growth potential simply because there is less mass for them to feed upon.

Step 3 focuses on the internal structure of all holes, and contrary with what we might expect, here we learn that despite what our imaginations may argue the interior of every hole must be blindingly illuminated! After all, the hole is absorbing light, not extinguishing it.

Step 4 takes in the effect of a toroidal universe which not only differentiates the photes of the ylem from those of the smoke ring but serves as a substratum through which the smoke ring drifts. The internal rotation of the smoke ring is the dynamic that determines the twist shape of the individual photes within it.

As for the photes themselves, they are even more minute than neutrinos, with the latter likely consisting of two photes united with their magnetic poles aligned to neutralize any element of charge.

Since each individual torus must conform to the direction of the eruption originating in the ylem, an aggregate total of six distinct and well identifiable universes are each characterized by its own peculiar degree of freedom, to which we add a seventh, straight line orientation which is independent of the torus but permeates it just as the air around a smoke ring contains the ring while it remains free of it.

Step 5 takes into account the consequences of a toroidal universe. Since every motion must follow a slightly curved track and the individual photes comprising the universe we happen to call 'home' there must be a bias imposing a rotation on every unit of matter confined by the torus hence every hole must be busily rotating, with a limiting rate of speed of C at its perimeter. It is not necessarily rotating this swiftly, but it must be spinning at a rate of at least several thousand km/sec. which is a huge number for an object as relatively small in size as a hole. This means the hole must be discoid in shape... a rotating flying saucer if you wish to think of it in those terms.

As an incidental aside not otherwise examined here, there is at least a remote possibility, given the curved structure of a toroidal universe, that our telescopes actually contain images of our own galaxy at one or more stages of its evolution!

To repeat, this is only a highly unlikely possibility, but it is one which can provide grist for the science fiction mill.

Step 6 focuses on hole mass increases. Obviously any random particles which chance to be passing by will add to the mass. This includes entire stars as well as stray protons or even electrons. But a significant source of new mass derives as a result of Olber's paradox. In the general core region there is no such thing as a nightfall and the sky is necessarily incandescent with the radiance of millions of stars flaming within it.

The result of Olber's paradox is the manufacture of a myriad stars pulsating wildly in an effort to expand and contract simultaneously; expanding to release excess heat while having to contract to produce heat needed to stoke its internal fiery reaction so it can delay its own collapse.

It is fundamentally an impossible situation which must fail over the long run. But it may work for a few million years before any one star succumbs and perhaps joins daddy in the hole. During this period the aggregate kinetic mass of the radiance being absorbed by the hole may have doubled or even tripled its gross mass from this source alone.

Step 7 outlines how a hole loses mass. Indescribable chaos characterizes the inner conditions within the surface and the transport time required for any given phote to move down from the hole surface to its core may exceed many thousand years of the classic 'drunkard's walk.

The chief loss of mass derives from the 7th or straight line phote configuration which is alien to this or any universe save the ylem megaverse As such it forms the substratum through which the torus moves. In other words, the straight line phote is the equivalent to the atmosphere through which the torus is moving, ergo any phote-to-phote collisions occurring within the hole which succeeds in deforming a curved phote into a straight line phote effectively drops it out of this universe and returns it to the ylem.

In this it may coincide with the 'worm-holing' escape first proposed by Hawking in which event it is not novel with us here.

There is also a potential escape route for photes deformed into configurations native to the universe from which they are derived, but this too is largely speculative and the likelihood of many escaping the hole by this route is slim.

Either way, there will be a literal Niagara of straight line photes tunneling their way out of the hole and escaping back into the ylem.

But how do they manage it? Here the answer is not entirely obvious. The mysterious 'dark matter' of our universe consists of the substratum of unwaved, i.e. invisible, photes through which the toroidal universe is moving. They are invisible precisely because our instruments are designed to detect waves and these photes have no wave length so we can only detect them by inference.

Step 8 considers the fate of holes developed outside the core region of a galaxy. Simple calculation argues that there must be several million of these in our galaxy alone, but they do not appear to be anywhere around. This would seem to be a fatal objection to all our models but it is easily overcome when we accept the idea of tunneling by any straight line photes back into the ylem from whence they originated.

Simply stated, as the hole is continually losing mass via this escape route the confining gravity effect (regardless of any questions of origin or notionality} is continually decreasing. Hole strength is declining and will continue to do so until the internal pressures demolish the barrier and we see a faint blue star which is adrift among the ordinary type 2 stars strewn about the galaxy.

The abnormal blueness is convincing evidence that it is one hot customer and the occasional high luminosity incandescent bursts emitted from these stars are persuasive arguments for the idea that they denote the final extinction of a hole and return to the more prosaic career of being merely a dead star.

Step 9, the final step, considers the environment of the core region as a whole. Here we encounter a region nearly as strange as we met within the hole itself. Radio telescopes pointed at the galactic core are filled with roars, thunderings, explosions, eerie whimperings and whining all of such titanic intensity we can still hear them 30,000 light years away!

These are not mere token sounds. They are stark evidence of chaos incarnate and a region where Brahma and Shiva are locked in deadly embrace, with one aspect manufacturing a host of exotic elements ranging upward to transuranic atoms while the Shiva aspect busies itself with induced electromagnetic work which expels the new creations either away from the hole region or into it. It is a veritable witches' cauldron of wildly competing energies, each striving and arguing fiercely with all the others for pride of possession but it is external to the hole rather than an emission from it. The hole itself produces nothing apart from the inevitable torrent of straight line photos.

These nine steps give us a deeper insight into the physical nature of our little galaxy and I believe it could not be achieved by any other approach than the one selected here.

Routine techniques provide only numbers without any help in understanding, but with understanding arriving first the numbers have a shot at coming up with those 1+1=2 level solutions so beloved by the bean counters.

And now we have at last reached our goal and can finally collate our steps and see where they lead us.

The results surprise even me by their clarity and novelty. I believe they may even be capable of disconcerting a few of the inevitable mocking birds who preen themselves in their ability to echo the notes of other birds.

But that is for others to decide.

CHAPTER VI

COMMENCEMENT DAY

Things have a way of turning out very differently than planned. As Bobby Burns penned in his Ode to a wee mousey, "The best laid plans o' mice and men oft gang aglay." When I set out to write Book 2 I had the naive idea I had resolved the main issues and would merely be putting the house in order.

What happened?

Instead of a few consolidating paragraphs outlining the role of black holes I found myself enmeshed in a quagmire of wholly unexpected chaos where I had somehow conceived of holes as something like basement archives filled with photes all neatly laid out like inert bodies; ghosts in a dismally dark mortuary doing nothing save exuding gravity.

The few added paragraphs regarding spiral arms were planned to be the capstone to add the finishing touches to our understanding of the universe and our galaxy.

All this came from existing literature, both formal as in the Astronomical and Astrophysical journals and college level textbooks which seemed extraordinarily reluctant to introduce ideas of life and motion into their science.

What did I find instead? I began with the plan of describing a condition but slowly came to accept the idea that I was not dealing with a condition but instead was being compelled to work out an interlocking system where no simple answers were to be found.

Where did I go that far astray? What factor or factors had I overlooked? It required over two months of carefully scrutinizing my logic step by step before I saw my error. It lay literally at my feet where I assigned two classifications to galaxies in our little universe!

There are not two classes; there are four! And that ought to have been the tip-off to the coming complexities.

Try it again and see what happens. Four types of galaxies, the largest class being the big spirals. Then we have a lesser class, the globulars. The third class are the irregulars, which appear to have made an effort to coalesce into ordinary galaxies but somehow failed in the attempt. Then lastly there are the Frebel globular clusters, which appear to have formed outside of some galaxy but lacked the material to resist incorporation by some nearby major galaxy.

So what do we have? We have automatic complexity plus a trigger to suggest there may be outside factors playing around with us.

Revisit our 9 steps focusing with the first few but also with visits to the other steps as needed.

Start with the expulsion of photes from the ylem, but this time we examine it more closely. In Book 1 we suggested that there was no big bang but instead there were hundreds of little bangs. There we allowed the matter to drop. Now we must refine our statement because it contains implications I overlooked in my initial conclusion.

Duns Scotus got into trouble with his superiors in the Church hierarchy when he ventured to say "Even God cannot make a triangle with four sides."

But suppose Scotus was correct and certain laws of nature cannot be violated with impunity? I have already introduced Kolmogorov into the primal ylem. Is there any reason I cannot throw in the old adage "Nature abhors a vacuum." and add it to the mix?

Try this and see what happens.

A more localized, smaller Kolmogorov cell develops and is promptly expelled into a forming universal creative matrix. Hard on its heels the ambient photes still within the ylem rush in to fill the void left by the expelled photes.

The detonations must still possess the six degrees of freedom required of a gaseous detonation in space (another echo of Duns Scotus and his dictum) so each explosion will possess the characteristic torque offered in Book 1, but this time we are able to add hiccups to the eruptions. Each explosion will be developed in relatively short order and will comprise the quantity of torqued photos to manufacture a galaxy within the newly formed universe.

But the process is pretty random and the various eruptions are developing over a large area. I earlier mentioned how a ranking prelate of the Roman Church refused to peer through Galileo's telescope and see the sunspots, uttering the curt remark that "I would rather believe a telescope would lie than believe God's holy creation would have the measles!" I greatly fear my remarks here would give the worthy gentleman apoplexy when I propose that the eruptive region is vast and the instability of the photos would result in chain reaction eruptions spewing galactic sided masses randomly through each of the up to six forming universes.

The majority of these expelled galaxies would have the wherewithal and density distribution to manufacture spirals, but some will lack the proper density distribution and thus be lacking the structure to develop core holes and thus spirals. Others will be spat out in a disorderly manner to and result in those irregular galaxies we see in space. Finally, when things are starting to settle down we see the appearance of those odd globular Frebel clusters.

In this we can detect a systematic progression where the initial galaxies arise from high density phote concentrations in a given region of the ylem and are generally spirals. Once the ylem phote density begins to thin out in any given region the newly generated galaxies tend to lack the necessary core density and we find the huge globular galaxies. Reduce the available photes even further and we arrive at irregular galaxies, perhaps like the Magellanic clouds. And lastly we see a few post prandial belches of Frebel clusters as coda to our entry into this universe.

This is our proper starting point in seeking to comprehend the universe. There can be no other.

Next we consider the character of the core hole as it is modified by our shift in choosing a starting point.

As already noted, each such hole will be characterized by a rapid rotation which may run into thousands of revolutions per second. Here Galileo's kinetic energy enters into the picture and the hole tends to pancake and bulge at the waist.

It may be of interest to point out that the orientation of the pancaked hole will ordinarily reflect the structure and orientation of the parent universe but this is not strictly necessary and a few extraneous factors such as adjacent galaxies, etc. may enter into the scenario. But one thing is certain; the highly flattened equatorial bulge will act the same as an ordinary fireworks pinwheel in vectoring the escaping straight line photes, which are also known as 'dark matter' to the mystification of those bean counters who bother to think on occasion.

The hole will therefore be aligned at a right angle to the plane of the galaxy from which vantage point it will be busily spraying the surrounding core stars with the fleeing dark matter.

Most of them will probably perfect their flight and rejoin the flock of background ylem photes through which our universe is drifting. But a significant percentage will interact with the curved photes streaming from the ambient core stars to commence the process of synthesizing elements.

In this it is noteworthy that the actual act of creation is not a product generated by the hole. The hole is merely a facilitator which delivers the raw material to the enveloping star field. The actual work takes place in the midst of the chaotic radiation and pulsations of the spaces between the stars and is a direct result of the partially effective Olber field which creates the chaos.

The mysterious thumpings, groans and wheezes invariably associated with received noise in the radio telescopes from a distance of some 30,000 light years provides ample proof that this is the seed bed of elemental creation so no more need be said on that score.

This brings us to the spiral arms conundrum. Why the spiral arms in the first place? How did they get started and how have they been maintained over a period which has lasted a bare minimum of 9 billion years, and possibly much longer even than that?

Why am I so positive of the minimum time scale? Think on it from a rational standpoint. Put the numbers together and see what happens. All the evidence of astrogeology and ordinary terrestrial geology converges on a system formation starting about 4.5 billion years ago. Since our system manifestly belongs to the Type 1 class the source material must have been ejected from the core star region at least that long ago.

Add to this the fact that the spiral arm continues a full 20,000 additional light years beyond our sun's circumgalactic orbit and we arrive at a potential 9 billion years as a minimal age for the spiral. It may even be greater than this.

It is far from precise but it is a valid working hypothesis.

As for the arms themselves, this is a direct product of the kinetic torque imposed on already charged exudates being cast away from the hole. In the absence of spin these particles and elementary atoms would be discharged roughly equally in all directions but here the powerful spin and discoid shape of the hole ram the fleeing photos squarely into the chaotic realm of the ambient core stars which manufactures an overwhelmingly powerful outbound electromagnetic acceleration akin to but immensely more potent ancestor to humankind's maglev railroad tracks.

Sorry if I sound like a typical crackpot when I continually declaim loudly of conditions encountered. Words such as 'immensely', 'powerful', 'titanic', etc. are not my native speech, but how else can I describe the conditions here when we speak not merely of forces which could gobble up the Earth without even belching but then throw in the sun and other members of the solar system for good measure?

I doubt if it can be done.

The fact that Population 1 stars are essentially confined to the spiral arm regions of the galaxy provides proof that the spiral arms do not represent outside matter being sucked into the core by the hole. That notion is absurd on the face of it. Where would it obtain the exotic heavy elements, the iron, the nickel the mercury and gold, etc.? From intergalactic space perhaps? And even if we postulate that it did we would logically expect to find evidence of Population 1 stars strewn above and below the plane of the system; which is utterly lacking. So forget that idea; it is born of invincible ignorance not of overly computerized science.

Population 1 stars must be the product of spiral arms. By extension we may retrace the careers of this Population 1 stuff as it departs the core.

Telescopes have revealed several galaxies which have expelled mysterious jets of energy at right angles to the plane of their galaxy. These jets are telescopically visible at distances ranging upward of 10,000 light years from their origin and they remain coherent even at that range.

If a solitary galaxy can generate that sort of energy should we be surprised at the idea our galaxy can emulate it? The real question here is why do these occasional events crop up in the first place?

Here I am frankly speculating and lack confidence in my wonderings but is it possible one of Frebel's minigalaxies may have collided head-on with the core of the anomalous galaxy? I imagine the collision of possibly 75,000 stars battering a core region would be enough to discommode it a bit, but I have no intention of pursuing it the matter anyhow though the thought is interesting to say the least.

In Book 1 I cited the shrunken quarter obtained through Edmund Scientific. Again quoting at some length, an ordinary clad alloy U.S. quarter is placed in a momentary magnetic field which is capable of exerting a force of around 100,000 amps throughout the coil. The coil configuration generates a current of roughly 1,000,000 amps within the quarter, which is comparable to the total electrical power consumed by a medium sized city.

In the case of my quarter only a relatively minor amperage was employed, running around the 60,000 mark since under extreme conditions both the apparatus and the quarter simply explode with an abrupt BANG! Tiny micropellets of shrapnel, glowing a blinding blue-white fly in every direction at speeds in excess of 5,760 km/sec. The whole affair is over in a quarter of a second.

Now we shift gears and visualize one of our quarters under the conditions existing among the core level stars. All of our fancy laboratory work would be merely a flea bite when stacked against the magnitude of the forces at play when a freshly brewed chunk of matter is cast into the exit chute of a spiral arm and heads outward to fulfill whatever fate destiny may have in store for it.

The conclusion is obvious. Despite all the efforts of musty pundits to trivialize the reality by converting it to tiny little squiggles on sheets of paper or bits of symbols in computers the fact remains; to truly comprehend the universe we must first attempt to grasp the magnitude of the forces at work. People with minuscule imaginations merely pontificate. Others with broader imaginations but scanty knowledge of physics are left in bewilderment. It requires a fusion of imagination and knowledge before we can realistically hope to understand the universe in which we live.

But this says nothing about the workings of the spiral arms and here we come face to face with a series of highly interesting conclusions. For one, the ejecta of spiral arms must consist of a highly charged gas plasma moving outward at barely sub-light speed while liberally admixed with both dark matter and the nuclei of heavier atoms, with the latter being formed from interactions with occasional dark matter photes along the way.

A secondary activity within the spiral is magnetic linkage of such nuclei as nickel, iron, or other similar elements. What began as a solitary atom or two of iron, for instance, will acquire literally tonnes or even megatonnes of mass while travelling out along the spiral. In the process of agglomeration a host of the familiar Population 1 stars and planetary systems will be forming as the increasingly potent gravitational emissions serve to scavenge out the surrounding debris. This leads to the surprising conclusion that probably every Population 1 star and planetary system was born within a spiral arm!

If this is true then why don't they stay safely tucked within their birthplace?

This turns out to be one of the relatively few places where the answer is simple. Larger particles such as planets and stars are bulky critters with a lot of inertia. The spiral arms were emitted around the periphery of a discoid shaped hole which was spinning around at a speed of several thousand km/hr. The lateral speed of the excreted photes necessarily matched the rotational velocity of the hole while the bulk of the newly aggregated matter was consistent with the normal orbital velocity inherent for objects at a distance from the core. In brief, the spiral arm simply outran the bulkier Population 1 material it had created and which was now in orbit about the core.

This presents a wholly unexpected possibility relevant to our own solar system and Earth. Geological evidence argues for a series of mass extinctions occurring at roughly 60 million year intervals going back to the earliest history of the planet. This pretty much coincides with eras when one or the other of the spiral arms would be passing through the solar orbit around the galaxy! As an added lagniappe this would typically constitute a period uniquely prone to meteor bombardments, including those which continue to scar the Earth as fossil craters.

It seems logical to attribute some of this to passages of the spiral but data to prove or disprove such a hypothesis is lacking, largely because no one has bothered to look into the possibility.

Shades of Wegener and continental drift!

Even if the earlier hypothesis is incorrect we still have enough meat to arrive at an unexpected but not irrational new conclusion. The observation that the spiral arms are incubators of newly hatched stars permits us to posit a size line of ejecta ranging from an ultra massive Sirius down to tiny pebbles as results of a spiral arm passage. There is nothing deliberate about it but if correct there ought to be a slightly higher concentration of type O stars nearer the galactic core than farther out beyond our sun's orbit. This would not be unexpected but if correct the model would also call for a generally regular progression in the distribution of debris in interstellar space.

Forget about Type B stars. They are all old and in the final phase of life before exploding, but the remainder of the stellar classification system is intact: O,A,F,G,K,M,R,N,S,I all designate sizes and luminosity extending far down to a barely radiating planet we call Jupiter, where the internal dynamics cause it to glow softly in the infra-red even when not irradiated by the sun.

Logic demands that interstellar space must also include all sizes and shades of brown bodies along with the other debris. Like it or not there will be eternally frozen Earths drifting along with no sun to warm them. There will be moons and asteroids strewn along the way. meteor swarms will abound. Stray pebbles will be commonplace.

In keeping with my earlier conclusions there will be no sense of order about it. It is all purely a random cast of the dice affair. I can only predict with confidence that interstellar space will be filled with all sorts of frozen surprises.

Whether any of this will carry over into intergalactic space is another matter. It may or may not.

It would be wrong of me to closet my remarks concerning spiral arms without mentioning a few systems termed 'ringed galaxies' because they form a complete circle around the perimeter while lacking any connecting spokes with the core region. Here again I am indulging in sheer speculation but I suspect this may lie in the future of all spiral galaxies, including our own.

Think of the implications here. It

is obvious we have arrived at the empty tank solution. For whatever reason the core region has finally run out of steam and is no longer fueling the plasma jets with newly manufactured heavy elements. There is nothing to sustain them and they gutter out. Whatever was on the way when the fuel ran out simply continues along its trajectory where it serves as a slowly decaying halo for the galaxy.

    Such will one day be the fate of our own galaxy.

CHAPTER VII

THE END

In writing this I have spanned some 13.5 billion years, starting with a primal ylem and continuing to the present. Though unorthodox, I believe I have accomplished my self-appointed task more accurately and certainly more completely than any other human ever has. And I have done so without hiding behind tidbits of theological or mystical elements to contaminate the picture.

If this be sheer arrogance then so be it. At worst it is a coherent depiction and not a patchwork quilt consisting of odds and ends the way so much of our contemporary astrophysics conveys it.

But there is still one task remaining. 13.5 billion years of past history deserves to be balanced a bit by a peek into the most likely remote future. When and how does it all end?

Two fundamentally different scenarios are currently favored by academicians. There is a yo-yo version where the universal expansion slows, stops and then collapses back to the initial big bang state so it can start all over and hope to get it right this time. The other version is more dismal. The stars and galaxies expand ever deeper into the void and slowly gutter out as the last of their nuclear fuel is exhausted. Ultimately we are reduced to no more than a cinder universe devoid of shape or form.

A third alternative merits mention even though it has been repudiated by its authors. This one held center stage during the 1950's and 60's and known as the Hoyle, Bondi and Gold steady state model. According to them the universe has been eternally the same overall though not the same in detail. As matter recedes over the observational horizon it is replaced by the appearance of new hydrogen atoms in the local interstellar space. The net quantity of matter in the visible space is forever fixed so for every subtraction there is an addition.

A brief analysis indicates that for the yo-yo model to work universal space must be highly structured and function rather like a child's balloon. Each universal incarnation inflates it to the limit before deflating it to start over. Any effort at manufacturing an open-ended yo-yo where the mass falls back fails to take into account the need to deduct the kinetic energy of the light which is left to travel through the otherwise void regions beyond the limits reached by the more material ingredients of our space. Each successive incarnation is therefore fated to begin with a lesser energy reserve than the one before it. This implies a given start-up date for the first incarnation and a terminal date when the newly collapsed universe will simply lack the resources to hiccup a replacement universe.

We lack any rational reason to advance the assumptions to sustain this model without recourse to arcane theology so I choose to discard it.

The dismal alternative where the infinite is populated by the cinder husks of burned out stars and frozen planets receding ever deeper into the void cannot be specifically refuted and thus must be admitted as a valid alternative.

Hoyle and his trio ultimately recanted the steady state model after two decades of intensive search failed to reveal so much as a hint of support. I can only accept their conclusion along with the general remark that we should always doubt any ad hoc solutions for problems if they require exotic conditions to sustain them. But there is a fourth alternative which does not require any exceptions to ordinary mechanics for it to work. To illustrate, consider that the focus of our interest has ever been centered on the universe of which we are a part, yet all of these endings have focused on stars or galaxies.

Rather than concern ourselves with such trivia as galaxies why not peer at the smoke ring which defines our universe? How does this compare with the fate of smoke rings puffed out in our parlors? The latter are merely microminiatures of the enormous smoke rings adrift in the ylem.

If a smoke ring was puffed out into a total vacuum devoid of all content. Even gravity, it would persist indefinitely, neither growing nor shrinking. But interference with outside molecules does not slow its motion or change its vector. Instead this outside impedimenta causes it to expand in dimensions without materially increasing the molecular count.

The smoke ring is therefore fraying about the edges. It is literally falling apart almost from the beginning of its career! This decay proceeds from the outside in until the final wisps of smoke are dissipated.

Translate this depiction upward to encompass the universe we inhabit and the conclusion is simultaneously obvious and outright stunning. It is trudging along through the megaversal ylem atmosphere, steadily growing in physical dimensions, steadily losing mass and coherency and steadily disintegrating.

Our universe is even now in its death throes!

But this does not affect humankind. The greatest loss is centered on the outer fringes of the universe and will likely require a few billion years before it affects our galaxy and our solar system.

This is not altogether a necessary conclusion and in remote theory it could happen tomorrow, but this would be tantamount to winning the national lottery ten times in a row!

Yes, it could happen, but it won't. If it has lasted a minimum of 13.5 billion years it will most likely hang on another 5 or 10 billion years so I doubt if any of us need worry about it.

And on this note I write,

FINI

A FEW NOTES

The lights grow dim as my blindness progresses. At nearly 90 years of age I have about reached the limit of my physical resources. The end of a journey which began some 70 years ago around the end of World War II, when I acquired a bound copy of Chandrasekhar's "Introduction to the Study of Stellar Structure" approaches. It took me nearly six years before I finally taught myself to understand what he said. It was also a painful learning experience because I felt it was needful to walk that path alone.

How well I succeeded is for others to judge; assuming anyone can be bothered.

I know I have made mistakes but it should be obvious I have no knowledge where or what they are. After all, if I knew these things I would have corrected them; which sometimes seems to be a rare talent on my part. So many people in power merely explode in fury and persist in denials when someone dares correct them. But then so many people who lack power are incapable of any degree of forgiveness when their leaders turn out to be human and venture to make mistakes. So if anyone decides to try a few potshots in my direction I bid them welcome, and merely request them to shoot true.

I had originally meant to throw in a few of my earlier publications dealing with solar systems, primarily our own.

Somewhere along the way it struck me that this would be primarily ego puffery and self-aggrandizement, so I content myself with noting that each of them focused on an aspect of astronomy which had hitherto been either ignored or overlooked by professionals in the field.

My sole resentment from this era resulted in consequence of a contact with the Jet Propulsion Laboratory in California. Since it illustrates the futility of any effort to communicate with academics who already know that whatever they believe is true must therefore be true, I will content myself by recounting the essentials without bothering to copy the article.

The time was about three years before the first fly-by of the planet Uranus. I entitled the article "The Mystery Planet" and it appeared in one of the numerous science fiction magazines which were so popular in that era.

Uranus is anomalous in our solar system since its rotational axis is tilted some $98°$ from the plane of the system, making it technically a retrograde motion. It also possessed a then known total of five satellite moons, all in nice circular orbits and all neatly aligned along the planetary axis.

No one could account for this oddity until a well qualified astronomer proposed the existence of a mystery dark planet in an extremely eccentric orbit centered out in the far reaches of the solar system but with a perihelion which brought it near Uranus every few hundred thousand years.

This approach caused Uranus to tilt on its axis and thus caused the peculiarity we see today.

It was such an obviously absurd proposition I could hardly comprehend how it gained even a smidgeon of notice, let along gain traction in literature. It was so ridiculous it was almost Congressional in magnitude. But at least Congress has an alibi for its ineptness. It consists of over 500 men and women busily scheming up ways of collecting money in order to stay in power, squabbling with their counterparts in the opposing party, bestowing their prejudices on the public and generally behaving like spoiled brats it is not to be wondered at that the result can be likened to the village halfwit.

But no excuse can apologize for the ridiculousness of the Nemesis hypothesis! If Uranus had been alone in its orbit it might perhaps be tolerated but there no known or even hypothetical way a Nemesis planet could possess the gravitational energy needed to tilt Uranus some 98^0 while simultaneously dragging all five moons along into orbits girdling the planet. At the very least it would strip them out of their neat little orbits and scatter them every which way. Worse yet, it would also have to distort the orbit of Uranus itself yet Uranus tracks one of the least eccentric orbits of any body in the solar system! In my article I concluded that the inclinational distortion was the product of a slight misalignment and overlapping of the Weizsacker cells.

Weizsacker was a brilliant astronomer and theoretician who possibly deserves a better reputation, but he made two mistakes which spoiled his chances. Worst was his misfortune in being an early supporter of Hitler and was a recipient of one of Himmler's cute little ploys of awarding useful individuals with honorary memberships in the SS. Weizsacker was duly appointed to the rank of SS Colonel. His membership dues were paid for by Himmler and he was forever tarred with the brush of Nazism and after that he was pretty much a persona non grata to the outside world.

Had he been a nuclear scientist, a rocket or tank designer or jet airplanes our intelligence goons would have slipped him into the country with no problems. But as a mere astrophysicist he fell well below the propaganda radar and vanished. After all, this was before the space race got underway.

But this is not the crux of my resentment of JPL. That came later, when an acquaintance sent me a copy of the advance protocol of the Uranus mission. It was well done and scholarly enough to impress me. It made no mention of Nemesis and about the only point I could add came as a result of my later study of the theoretical magnetic fields of a spinning Uranus as they related to the solar magnetic flux.

These indicated that the magnetic field of Uranus would have to be either polar or displaced 98^0 but it could not be both. So far. so good. I wrote them thanking them for sending the copy of their protocol and added that they ought to expect a displacement of some 45^0 in the magnetic field and adding that lacking knowledge of the parent orientation of Uranus I could not predict whether the 8^0 loophole would be either plus or minus.

There the matter rested until the fly-by a couple of years later, followed in short order by a publicity blurb to the effect that to the amazement of the researchers the magnetic field of Uranus was offset by around $45°$ and nobody had ever suggested such a thing!

Evidently I was the nobody who had suggested it. I lacked the cachet to venture onto their sacred turf with profane feet. I have carried this resentment on my back for lo these many years.

I shall probably die with it. Gratuitous arrogance may be tolerated in politicians, in lawyers and with bankers or business tycoons, but it ought never to be tolerated in academia. Knowledge from whatever source remains knowledge and must be accepted on its own terms. But it seldom is.

70 odd years of work are now complete.

THE FINAL END

www.ingramcontent.com/pod-product-compliance
Lightning Source LLC
Chambersburg PA
CBHW051920170526
45168CB00001B/467